New Developments in Functional and Fractional Differential Equations and in Lie Symmetry

New Developments in Functional and Fractional Differential Equations and in Lie Symmetry

Editors

Ioannis P. Stavroulakis
Hossein Jafari

MDPI • Basel • Beijing • Wuhan • Barcelona • Belgrade • Manchester • Tokyo • Cluj • Tianjin

Editors
Ioannis P. Stavroulakis
University of Ioannina
Greece

Hossein Jafari
University of South Africa
South Africa

Editorial Office
MDPI
St. Alban-Anlage 66
4052 Basel, Switzerland

This is a reprint of articles from the Special Issue published online in the open access journal *Symmetry* (ISSN 2073-8994) (available at: https://www.mdpi.com/journal/symmetry/special_issues/New_developments_Lie_Symmetry).

For citation purposes, cite each article independently as indicated on the article page online and as indicated below:

LastName, A.A.; LastName, B.B.; LastName, C.C. Article Title. *Journal Name* **Year**, *Volume Number*, Page Range.

ISBN 978-3-0365-1158-0 (Hbk)
ISBN 978-3-0365-1159-7 (PDF)

© 2021 by the authors. Articles in this book are Open Access and distributed under the Creative Commons Attribution (CC BY) license, which allows users to download, copy and build upon published articles, as long as the author and publisher are properly credited, which ensures maximum dissemination and a wider impact of our publications.

The book as a whole is distributed by MDPI under the terms and conditions of the Creative Commons license CC BY-NC-ND.

Contents

About the Editors vii

Preface to "New Developments in Functional and Fractional Differential Equations and in Lie Symmetry" ix

Nematollah Kadkhoda, Elham Lashkarian, Mustafa Inc, Mehmet Ali Akinlar and Yu-Ming Chu
New Exact Solutions and Conservation Laws to the Fractional-Order Fokker–Planck Equations
Reprinted from: *Symmetry* **2020**, *12*, 1282, doi:10.3390/sym12081282 1

Subhadarshan Sahoo, Santanu Saha Ray, Mohamed Aly Mohamed Abdou, Mustafa Inc and Yu-Ming Chu
New Soliton Solutions of Fractional Jaulent-Miodek System with Symmetry Analysis
Reprinted from: *Symmetry* **2020**, *12*, 1001, doi:10.3390/sym12061001 15

Nichaphat Patanarapeelert, Saowaluck Chasreechai and Thanin Sitthiwirattham
On Nonlinear Fractional Difference Equation with Delay and Impulses
Reprinted from: *Symmetry* **2020**, *12*, 980, doi:10.3390/sym12060980 29

Afshin Babaei, Hossein Jafari and S. Banihashemi
A Collocation Approach for Solving Time-Fractional Stochastic Heat Equation Driven by an Additive Noise
Reprinted from: *Symmetry* **2020**, *12*, 904, doi:10.3390/sym12060904 43

Emad R. Attia, Hassan A. El-Morshedy and Ioannis P. Stavroulakis
Oscillation Criteria for First Order Differential Equations with Non-Monotone Delays
Reprinted from: *Symmetry* **2020**, *12*, 718, doi:10.3390/sym12050718 59

B. Muatjetjeja, S. O. Mbusi and A. R. Adem
Noether Symmetries of a Generalized Coupled Lane-Emden-Klein-Gordon-FockSystem with Central Symmetry
Reprinted from: *Symmetry* **2020**, *12*, 566, doi:10.3390/sym12040566 77

Eyaya Fekadie and Zhoushun Zheng
Finite Difference Approximation Method for a Space Fractional Convection–Diffusion Equation with Variable Coefficients
Reprinted from: *Symmetry* **2020**, *12*, 485, doi:10.3390/sym12030485 83

Ábel Garab
A Sharp Oscillation Criterion for a Linear Differential Equation with Variable Delay
Reprinted from: *Symmetry* **2019**, *11*, 1332, doi:10.3390/sym11111332 103

Áron Fehér, Lőrinc Márton and Mihály Pituk
Approximation of a Linear Autonomous Differential Equation with Small Delay
Reprinted from: *Symmetry* **2019**, *11*, 1299, doi:10.3390/sym11101299 113

Alexander Domoshnitsky, Irina Volinsky and Marina Bershadsky
Around the Model of Infection Disease: The Cauchy Matrix and Its Properties
Reprinted from: *Symmetry* **2019**, *11*, 1016, doi:10.3390/sym11081016 123

About the Editors

Ioannis P. Stavroulakis holds a master's degree in mathematics (City University of New York), a Ph.D. in mathematics (University of Ioannina) and a doctor honoris causa (University of Gjirokastra). His research interests lie in the area of Qualitative theory of differential equations; retarded and advanced differential equations; difference equations; functional equations; dynamic equations; partial differential equations. He is the author of 3 books and more than 140 research papers, most of them of *high quality*, published in *superior journals with excellent reviews* from the editors and/or referees, and cited very often (more than *3000 citations*). He has held positions at 10 universities in several countries and has been invited as a member of the *scientific/organizing* committee and/or as a *keynote/plenary/invited speaker* at many international conferences and universities, delivering *more than 130 lectures at 130 universities in 30 countries in the 5 continents* and collaborated with *more than 70 researchers* around the world. He is also *Editor-in-Chief, Managing Editor, Guest Editor, or Editor* of 14 mathematical journals. He was awarded several awards and distinctions among them: *Ampere Foundation Fellowship; Canon Foundation in Europe Research Fellowship; The Flinders University of South Australia Visiting Research Fellowship; DOCTOR HONORIS CAUSA* (he was the first to be awarded an Honorary Doctorate from the University of Gjirokastra). Two Special Issues have been dedicated to him: *Global and Stochastic Analysis* 5 No.1, (2018) (www.mukpublications.com) and *Nonlinear Dynamics and Systems* Theory 19 (1-SI) (2019) (www.e-ndst.kiev.ua). He received a certificate of OBADA-PRIZE 2019 (in the top 10 out of 370).

Hossein Jafari Professor Hossein Jafari is an accomplished applied mathematician. His research area is in fractional-order differential equations and their applications. According to the American Mathematical Society database, Prof Jafari's publications cover the following interesting topics:

- Numerical Analysis

- Real Functions

- Linear And Multilinear Algebra; Matrix Theory

- Calculus Of Variational and Optimization Control; Optimization

- Integral Equations

- Mechanics And Deformable Solids

- Mechanics Of Particles; And Systems

- Fluid Mechanics

- Classical Thermodynamics, Heat Transfer

- Quantum Theory

- Biology and Other Natural Sciences

In 2020, Prof. Jafari introduced a new general integral transform. As a result of his extensive research output and citations, he has been included on the list of the top 2% of scientists in the world as compiled by Stanford University (USA). Since 2017, he has also been on the list of the Essential Science Indicators (ESI) as a top researcher by the Web of Science, which lists the top 1% scientists in the world.

Preface to "New Developments in Functional and Fractional Differential Equations and in Lie Symmetry"

Ordinary differential equations (ODEs) appear frequently in mathematical models that attempt to describe real-life situations in which the rate of change of the system depends only on its present stage. However, in many cases, the past state of the system has to be taken into consideration. Delay differential equations or differential equations with retarded argument or hystero-differential equations provide more realistic mathematical models for systems in which the rate of change depends not only on their present stage but also on their past history, such as population models, models for epidemics, economic models, nuclear reactors, collision problems in electrodynamics, and many others. In recent years, there has also been a great deal of interest in the study of the discrete analogue difference equations.

Many physical phenomena in areas such as electrochemistry, physics, biology, mechanics, signal processing, and viscoelastic materials can be modelled using fractional derivatives. Fractional calculus is a generalization of differentiation and integration to arbitrary non-integer order.

The method of group analysis of differential equations was introduced by Sophus Lie more than one hundred years ago. A symmetry transformation maps an equation into itself. The set of such transformations forms a Lie group and gives rise to Lie algebra, which enables easier manipulation of differential equations. The Lie symmetry method is a powerful tool to solve or reduce ODEs and a way to find exact solutions of partial differential equations (PDEs) by reducing the number of independent variables in the equations and solve engineering and applied science problems, which are modelled in terms of nonlinear and complicated ODEs and PDEs.

In this Special Issue, recent developments on the above-mentioned areas are presented by experts on the subjects. The guest editors believe that the papers published in this Special Issue will be useful to a wide range of researchers and will motivate further research in the topics presented as well as in the related fields.

Ioannis P. Stavroulakis, Hossein Jafari
Editors

Article

New Exact Solutions and Conservation Laws to the Fractional-Order Fokker–Planck Equations

Nematollah Kadkhoda [1], Elham Lashkarian [2], Mustafa Inc [3], Mehmet Ali Akinlar [4] and Yu-Ming Chu [5,6,*]

[1] Department of Mathematics, Faculty of Basic Sciences, Bozorgmehr University of Qaenat, Qaenat, Iran; kadkhoda@buqaen.ac.ir
[2] Department of Mathematical Sciences, Shahrood University of Technology, Shahrood, Semnan, Iran; lashkarianelham@yahoo.com
[3] Department of Mathematics, Firat University, 23119 Elazig, Turkey; minc@firat.edu.tr
[4] Department of Mathematical Engineering, Yildiz Technical University, 34220 Istanbul, Turkey; makinlar@yildiz.edu.tr
[5] Department of Mathematics, Huzhou University, Huzhou 313000, China
[6] Hunan Provincial Key Laboratory of Mathematical Modeling and Analysis in Engineer, Changsha University of Science & Technology, Changsha 410114, China
* Correspondence: chuyuming@zjhu.edu.cn

Received: 6 July 2020; Accepted:29 July 2020; Published: 3 August 2020

Abstract: The main purpose of this paper is to present a new approach to achieving analytical solutions of parameter containing fractional-order differential equations. Using the nonlinear self-adjoint notion, approximate solutions, conservation laws and symmetries of these equations are also obtained via a new formulation of an improved form of the Noether's theorem. It is indicated that invariant solutions, reduced equations, perturbed or unperturbed symmetries and conservation laws can be obtained by applying a nonlinear self-adjoint notion. The method is applied to the time fractional-order Fokker–Planck equation. We obtained new results in a highly efficient and elegant manner.

Keywords: lie point symmetry analysis; approximate conservation laws; approximate nonlinear self-adjointness; perturbed fractional differential equations

MSC: 22E10; 35L65; 47A05; 26A33

---

1. Introduction

Fractional partial differential equations are a generalization of classical ordinary calculus with utilizations of integrals and derivatives with an arbitrary order. In the last decade, these equations were employed in various scientific and engineering phenomena including fluid mechanics, gas dynamics, nonlinear acoustics, biology, control theory, earthquake modeling, traffic flow models. There are several different types of fractional-order derivative and integral operators including the Riesz, Riemann–Liouville, Grünwald–Letnikov and Caputo fractional derivatives [1].

We are concerned with approximations using a small parameter of the Caputo and Riemann–Liouville type fractional derivative operators. Using this approximation, a fractional-order differential equation may be converted into an integer-order equation [2–7].

By the Lie symmetry techniques [8–10], we can obtain analytical solutions of many perturbed differential equations. Noether's theorem which was introduced by Emmy Noether in 1918 describing general concepts related to symmetry groups and conservation laws is a useful tool in the solutions of perturbed differential equations, see, e.g., [11–13]. Finding approximate symmetries of perturbed partial dofferential equations was first introduced by Fushchich, Shtelen and Baikov [14,15]. Because of the importance of perturbed systems to describe the natural phenomena, they generalized the Noether's theorem to approximated version. This generalization helps to find approximate conservation laws of a given system including the related topics [16,17]. For a system, approximate conservation laws is determined by approximate formal Lagrange and nonlinear self-adjointness for approximate equations [18]. We present conservation laws of fractional partial differential equations [19,20] with an effective method based on nonlinear self-adjointness.

The Fokker–Planck equations play an important role in fluid mechanics, control theory, astrophysics and quantum [21,22]. We are concerned with the perturbed fractional-order Fokker–Planck equation

$$D_t^\alpha u - \frac{1}{2}a^2 u_{xx} - bu - bxu_x + \varepsilon u_t = 0. \tag{1}$$

In which a, b are constants and D_t^α is fractional derivative of order α.

2. Approximation of Fractional-Order Operators

Definition 1. *The left and right-sided Riemann–Liouville fractional partial derivatives are defined as*

$$\left({}_a D_{x^1}^{\alpha+k} u\right)(x) = \frac{1}{\Gamma(1-\alpha)}\left(\frac{\partial}{\partial x^1}\right)^{k+1} \int_a^{x^1} \frac{u(\xi, x^2, \ldots, x^n)}{(x^1-\xi)^\alpha} d\xi, \tag{2}$$

$$\left({}_{x^1} D_b^{\alpha+k} u\right)(x) = \frac{(-1)^{k+1}}{\Gamma(1-\alpha)}\left(\frac{\partial}{\partial x^1}\right)^{k+1} \int_{x^1}^b \frac{u(\xi, x^2, \ldots, x^n)}{(x^1-\xi)^\alpha} d\xi. \tag{3}$$

Respectively in which $\Gamma(\cdot)$ denotes the Gamma function and $\alpha \in (0,1)$, $k = 0, 1, \ldots, m$, $m \in \mathbb{N}$.

Definition 2. *The left and right-sided Caputo type fractional partial derivative are defined as*

$$\left({}_a^C D_{x^1}^{\alpha+k} u\right)(x) = \frac{1}{\Gamma(1-\alpha)} \int_a^{x^1} \frac{1}{(x^1-\xi)^\alpha} \frac{\partial^{k+1} u(\xi, x^2, \ldots, x^n)}{\partial \xi^{k+1}} d\xi, \tag{4}$$

$$\left({}_{x^1}^C D_b^{\alpha+k} u\right)(x) = \frac{(-1)^{k+1}}{\Gamma(1-\alpha)} \int_{x^1}^b \frac{1}{(\xi-x^1)^\alpha} \frac{\partial^{k+1} u(\xi, x^2, \ldots, x^n)}{\partial \xi^{k+1}} d\xi. \tag{5}$$

Respectively in which $\Gamma(\cdot)$ denotes the Gamma function and $\alpha \in (0,1)$, $k = 0, 1, \ldots, m$, $m \in \mathbb{N}$.

For the natural numbers, k, c, d, let $u(x) := u$ be the function of $x = (x^1, x^2, \ldots, x^n) \in \mathbf{R}^n$, we consider an fractional differential equation in the form of

$$P\left(x, u, u_{(1)}, \ldots, u_{(k)}, {}_a D_{x^1}^{\alpha_0}, {}_a D_{x^1}^{\alpha_1}, \ldots, {}_a D_{x^1}^{\alpha_d}, D_{b,x^1}^{\beta_0}, D_b^{\beta_1}, \ldots, {}_{x^1} D_b^{\beta_c}\right) = 0, \tag{6}$$

$0 < \alpha_0 < \alpha_1 < \ldots < \alpha_d$, $0 < \beta_0 < \beta_1 < \ldots < \beta_c$.

The partial derivative of u is denoted as

$$u_{(s)} \equiv \{u_{i_1\ldots i_s}\} = \{\frac{\partial^s u(x)}{\partial x^{i_1} \ldots \partial x^{i_s}}\}, \qquad (i_1, \ldots, i_s = 1, \ldots n, s = 1, \ldots, k).$$

If the orders of fractional differential Equation (6) are all nearly integers, then it is possible to approximate Equation (6):

$$P\left(x, u, u_{(1)}, \ldots, u_{(r)}, {}_a D_{x^1}^{\alpha}, {}_a D_{x^1}^{\alpha+1}, \ldots, {}_a D_{x^1}^{\alpha+d}, D_{b,x^1}^{\alpha}, D_b^{\alpha+1}, \ldots, {}_{x^1} D_b^{\alpha+d}\right) = 0. \tag{7}$$

In which $\alpha \in (0,1)$. Assuming $\alpha = \varepsilon$ or $\alpha = 1 - \varepsilon$ in Equation (7), we can turn the right and left-sided Riemann–Liouville fractional partial derivatives into a Taylor expansion having arbitrarily small parameter, $1 > \varepsilon > 0$.

Supposing the existence of each derivative ${}_aD_{x^1}^{k+\varepsilon}u$, ${}_{x^1}D_b^{k+\varepsilon}u$ $(k = 0, 1, \ldots)$ or ${}_aD_{x^1}^{k-\varepsilon}u$, ${}_{x^1}D_b^{k-\varepsilon}u$ $(k = 1, 2, \ldots)$ at arbitrary point $x^1 \in (a,b)$, we have

$$\begin{aligned}
{}_aD_{x^1}^{k\pm\varepsilon}u &= \sum_{s=0}^{\infty} \binom{k\pm\varepsilon}{s} \frac{(x^1 - a)^{s-k\mp\varepsilon}}{\Gamma(1-k+s\mp\varepsilon)} \frac{\partial^s u(\xi, x^2, \ldots, x^n)}{\partial \xi^s} \\
&= \frac{\partial^k u}{\partial (x^1)^k} \pm \varepsilon \left([\psi(k+1) - \ln(x^1 - a)] \frac{\partial^k u}{\partial (x^1)^k} \right. \\
&\quad \left. - \sum_{s=0, s\neq k}^{\infty} \frac{(-1)^{s-k} k!}{(s-k) s!} (x^1 - a)^{s-k} \frac{\partial^s u}{\partial (x^1)^s} \right) + o(\varepsilon),
\end{aligned} \quad (8)$$

$$\begin{aligned}
{}_{x^1}D_b^{k\pm\varepsilon}u &= \frac{\partial^k u}{\partial (x^1)^k} \pm \varepsilon \left([\psi(k+1) - \ln(b - x^1)] \frac{\partial^k u}{\partial (x^1)^k} \right. \\
&\quad \left. - \sum_{s=0, s\neq k}^{\infty} \frac{(-1)^{s-k} k!}{(s-k) s!} (b - x^1)^{s-k} \frac{\partial^s u}{\partial (x^1)^s} \right) + o(\varepsilon).
\end{aligned} \quad (9)$$

Here, $\psi(z) = \frac{\Gamma'(z)}{\Gamma(z)}$ is the digamma function and $\binom{k\pm\varepsilon}{s} = \frac{\Gamma(1+k\pm\varepsilon)}{\Gamma(1+k-s\pm\varepsilon)s!}$ is a binomial coefficient.

For the Caputo fractional derivative

$${}_a^C D_{x^1}^{k\pm\varepsilon}u = {}_aD_{x^1}^{k\pm\varepsilon}u \mp \varepsilon \sum_{s=0}^{k-1} (-1)^{s-k}(k-s-1)!(x^1-a)^{s-k} \frac{\partial^s u}{\partial (x^1)^s}\Big|_{x^1=a} + p(x,a), \quad (10)$$

$${}_{x^1}^C D_b^{k\pm\varepsilon}u = {}_{x^1}D_b^{k\pm\varepsilon}u \mp \varepsilon \sum_{s=0}^{k-1} (-1)^{s-k}(k-s-1)!(b-x^1)^{s-k} \frac{\partial^s u}{\partial (x^1)^s}\Big|_{x^1=b} + q(x,b). \quad (11)$$

In which

$$p(x,a) = \begin{cases} -[1 + \varepsilon(\psi(1) - \ln(x^1 - a))] \frac{\partial^k u}{\partial (x^1)^k}\Big|_{x^1=a}; & \text{for } {}_a^C D_{x^1}^{k+\varepsilon} u, \\ 0; & \text{for } {}_a^C D_{x^1}^{k-\varepsilon} u, \end{cases}$$

$$q(x,b) = \begin{cases} -[1 + \varepsilon(\psi(1) - \ln(b - x^1))] \frac{\partial^k u}{\partial (x^1)^k}\Big|_{x^1=b}; & \text{for } {}_{x^1}^C D_b^{k+\varepsilon} u, \\ 0; & \text{for } {}_{x^1}^C D_b^{k-\varepsilon} u. \end{cases}$$

Proposition 1. *Let F be a continuously differentiable function with respect to ${}_aD_{x^1}^{\alpha+k}u$ and ${}_{x^1}D_b^{\alpha+k}u$ $(k = 0, 1, \ldots, d)$. Then, for $\alpha = \varepsilon$ or $\alpha = 1 - \varepsilon$, we can approximate Equation (7) as follows:*

$$P_{(0)}(x, u, u_{(1)}, \ldots) + \varepsilon P_{(1)}(x, u, u_{(1)}, \ldots, D_{x^1}^{c+1}u, D_{x^1}^{c+2}u, \ldots) \approx 0, \quad (12)$$

in which $c = \max\{d, r\}$ for $\alpha = 1 - \varepsilon$ and $c = \max\{d-1, r\}$ for $\alpha = \varepsilon$.

3. Lie Group Analysis

We consider a differential operator of first order defined as

$$X \approx X_{(0)} + \varepsilon X_{(1)}$$
$$\equiv \left(\zeta^i_{(0)}(x,u) + \varepsilon \zeta^i_{(1)}(x,u)\right)\frac{\partial}{\partial x^i} + \left(\theta_{(0)}(x,u) + \varepsilon \theta_{(1)}(x,u)\right)\frac{\partial}{\partial u}, \quad (13)$$

in which

$$\zeta^i_{(0)}(x,u) = \frac{\partial g^i_{(0)}(x,u,a)}{\partial a}\Big|_{a=0}, \quad \zeta^i_{(1)}(x,u) = \frac{\partial g^i_{(1)}(x,u,a)}{\partial a}\Big|_{a=0},$$
$$\theta_{(0)}(x,u) = \frac{\partial h_{(0)}(x,u,a)}{\partial a}\Big|_{a=0}, \quad \theta_{(1)}(x,u) = \frac{\partial h_{(1)}(x,u,a)}{\partial a}\Big|_{a=0}.$$

Calculating the solutions of

$$X\left(P_{(0)} + \varepsilon P_{(1)}\right)\Big|_{(12)} \approx 0, \quad (14)$$

exact symmetries of the perturbed Equation (7) can be achieved.

$$\begin{aligned}\tilde{x}^i &\approx g^i(x,u,a,\varepsilon) \equiv g^i_{(0)}(x,u,a) + \varepsilon g^i_{(1)}(a,u,a),\\ \tilde{u} &\approx h(x,u,a,\varepsilon) \equiv h_{(0)}(x,u,a) + \varepsilon h_{(1)}(a,u,a),\end{aligned} \quad (15)$$

with

$$\tilde{x}^i|a=0 \approx x^i, \quad \tilde{u}|_{a=0} = u,$$

are group of Lie point transformations under the group conditions

$$g^i\left(g^1(x,u,a,\varepsilon),\ldots,g^n(x,u,a,\varepsilon),h(x,u,a,\varepsilon),b,\varepsilon\right) \approx g^i(x,a+b,\varepsilon);$$
$$h\left(g^1(x,u,a,\varepsilon),\ldots,g^n(x,u,a,\varepsilon),h(x,u,a,\varepsilon),b,\varepsilon\right) \approx h(x,a+b,\varepsilon),$$

by $o(\varepsilon)$.

4. Classification of Group-Invariant Solution

We present the optimal system of approximate Fokker–Planck equation symmetries [23] by employing the fact that every s-dimensional subalgebra is equivalent to a unique member of the optimal system with an adjoint representation. If we know the infinitesimal adjoint action $ad\mathbf{g}$ of a Lie algebra $\mathbf{g}$ on itself, we can reconstruct the adjoint representation AdG of the underlying Lie group.

$$\frac{dX}{d\varepsilon} = adY|_X, \quad X(0) = X_0,$$

with solution

$$X(\varepsilon) = Ad\left(\exp(\varepsilon Y)\right) X_0,$$

where

$$Ad\left(\exp(\varepsilon Y)\right) X_0 = \sum_{n=0}^{\infty} \frac{\varepsilon^n}{n!}(adY)^n(X_0) = X_0 - \varepsilon[Y, X_0] + \frac{\varepsilon^2}{2}[Y,[Y,X_0]] - \ldots.$$

It is clear that $[X_i, X_j]$ is the usual commutator and ε is a parameter.

Optimal System and Exact Solutions

Consider the perturbed fractional-order Fokker–Planck equation

$$_0D_t^\alpha u = \frac{1}{2}a^2 u_{xx} + bu + bxu_x - \varepsilon u_t, \quad u = u(x,t), \quad \alpha \in (0,1). \tag{16}$$

In order to calculate the approximate symmetries of the perturbed fractional equation, we apply the extension of Equation (8) to Equation (16). Setting $\alpha = 1 - \varepsilon$, we can write Equation (16) as

$$\begin{aligned} P_{(0)} + \varepsilon P_{(1)} &= u_t - \frac{1}{2}a^2 u_{xx} - bu - bxu_x \\ &+ \varepsilon\left[(\ln t + v)u_t + u + \sum_{k=1}^{\infty}\frac{(-t)^k}{k(k+1)!}u_t^{(k+1)}\right] + \varepsilon u_t = 0. \end{aligned} \tag{17}$$

We get symmetries of perturbed equation Equation (17) using the Maple software.

$$\begin{aligned} X_1 &= \partial_t, \quad X_2 = u\partial_u, \quad X_3 = e^{-bt}\partial_x, \quad X_4 = e^{bt}\partial_x - \frac{2bxu}{a^2}e^{bt}\partial_u, \\ X_5 &= e^{-2bt}\partial_t - bxe^{-2bt}\partial_x + bue^{-2bt}\partial_u, \quad X_6 = bxe^{2bt}\partial_t + bxe^{2bt}\partial_x - \frac{2b^2x^2u}{a^2}e^{2bt}\partial_u, \\ X_7 &= e^{bt + \frac{c_1 t}{2} - \frac{b}{a^2}x^2}\operatorname{KummerM}(\frac{4b + c_1}{4b}, \frac{3}{2}, \frac{b}{a^2}x^2)\partial_u, \\ X_8 &= e^{bt + \frac{c_1 t}{2} - \frac{b}{a^2}x^2}\operatorname{KummerU}(\frac{4b + c_1}{4b}, \frac{3}{2}, \frac{b}{a^2}x^2)\partial_u. \\ Y_1 &= \frac{-\varepsilon}{a^2}\partial_t, \quad Y_2 = \varepsilon u\partial_u, \quad Y_3 = \frac{-\varepsilon}{a^2}e^{-bt}\partial_x, \quad Y_4 = \frac{-\varepsilon}{a^2}e^{bt}(\partial_x + 2bxu\partial_u), \\ Y_5 &= \frac{\varepsilon}{a^2}e^{-2bt}\left(\partial_t + bx\partial_x + ba^2 u\partial_u\right), \quad Y_6 = \frac{\varepsilon}{a^2}e^{2bt}\left(\partial_t - bx\partial_x - 2b^2x^2u\partial_u\right), \\ Y_7 &= x\varepsilon e^{tc_1 - \frac{bx^2}{a^2}}\operatorname{KummerM}(\frac{b + c_1}{2b}, \frac{3}{2}, \frac{b}{a^2}x^2)\partial_u, \\ Y_8 &= x\varepsilon e^{tc_1 - \frac{bx^2}{a^2}}\operatorname{KummerU}(\frac{b + c_1}{2b}, \frac{3}{2}, \frac{b}{a^2}x^2)\partial_u. \end{aligned} \tag{18}$$

where the Kummer functions, $\operatorname{KummerM}(\mu, \nu, z)$ and $\operatorname{KummerU}(\mu, \nu, z)$ solve the differential equation $zy'' + (\nu - z)y' - \mu y = 0$.

By the possession of infinitesimal generators (18), a number of adjoint representations are given as

$$\begin{aligned} &Ad[X_1, X_j] = X_j, \quad j = 1, \ldots, 5, \quad Ad[X_i, X_i] = X_i, \quad i = 1, \ldots, 5, \\ &Ad[X_2, X_1] = X_1 - \varepsilon b X_3, \quad Ad[X_2, X_4] = \frac{2\varepsilon b}{a^2}X_2 + X_4, \\ &Ad[X_3, X_1] = X_1 - \varepsilon b X_4, \quad Ad[X_3, X_4] = \frac{2\varepsilon b}{a^2}X_2 + X_4, \\ &Ad[X_4, X_1] = X_1 + \varepsilon b X_4, \quad Ad[X_4, X_3] = -\frac{2\varepsilon b}{a^2}X_2 + X_3, \\ &Ad[X_4, X_5] = -\frac{2\varepsilon^2 b^2}{a^2}X_2 + 2\varepsilon b X_3 + X_5, \quad Ad[X_5, X_1] = X_1 - 2\varepsilon b X_5, \\ &Ad[Y_1, Y_j] = Y_j, \quad j = 1, \ldots, 4, \quad Ad[Y_i, Y_i] = Y_i, \quad i = 1, \ldots, 4, \\ &Ad[Y_2, Y_1] = Y_1 - \frac{\varepsilon b}{a^2}Y_3, \quad Ad[Y_2, Y_4] = -\frac{2\varepsilon b}{a^4}Y_2 + Y_4, \\ &Ad[Y_3, Y_1] = Y_1 - \frac{\varepsilon b}{a^2}Y_3, \quad Ad[Y_3, Y_4] = -\frac{2\varepsilon b}{a^4}Y_3 + Y_4, \\ &Ad[Y_4, Y_1] = Y_1 + \frac{\varepsilon b}{a^2}Y_4, \quad Ad[Y_4, Y_3] = \frac{2\varepsilon b}{a^4}Y_2 + Y_3, \ldots \end{aligned}$$

Suppose that $V = \sum_{i=1}^{8} X_i$ and $\tilde{V} = \sum_{i=1}^{8} Y_i$ are the most general element. Eventually, we will obtain one-dimensional optimal system of Equation (18). The following symmetries are just a few members of optimal system of the perturbed Fokker–Planck equation

$$V_1 = X_1, \quad V_2 = X_2, \quad V_3 = X_3, \quad V_4 = X_4 \quad V_5 = X_2 + X_3,$$
$$V_6 = X_5, \quad V_7 = X_6, \quad V_8 = X_7, \quad V_9 = X_8 \quad V_{10} = X_0,$$
$$V_{11} = X_3 + X_5, \quad V_{12} = X_2 + X_5, \quad V_{13} = X_2 + X_4, \quad V_{14} = X_1 + X_4,$$
$$V_{15} = X_2 + X_3 + X_4, \quad V_{16} = X_2 + X_3 + X_5, \quad V_{17} = X_1 + X_2 + X_4,$$
$$V_{18} = X_1 + X_3 + X_4, \quad V_{19} = X_1 + X_4 + X_5, \quad V_{20} = X_2 + X_3 + X_4 + X_5, \ldots$$
$$\tilde{V}_1 = Y_1, \quad \tilde{V}_2 = Y_2, \quad \tilde{V}_3 = Y_3, \quad \tilde{V}_4 = Y_4 \quad \tilde{V}_5 = Y_2 + Y_4,$$
$$\tilde{V}_6 = Y_1 + Y_3 + Y_4, \quad \tilde{V}_7 = Y_1 + Y_2 + Y_3, \ldots$$

Case 1: For the symmetry of $V_1 = X_1$, corresponding characteristic equation is given as:

$$\frac{dt}{1} = \frac{dx}{0} = \frac{du}{0}, \tag{19}$$

integration of Equation (19) yields the following similarity variable and function

$$u = g(x), \tag{20}$$

thus we have

$$u_t = 0, \quad u_x = g'(x), \quad u_{xx} = g''(x). \tag{21}$$

Substituting Equations (20) and (21) into Equation (17), we can get the reduced equation:

$$-\frac{1}{2}a^2 g'' - bg - bxg' + \varepsilon \left[\frac{g}{t}\right] = 0,$$

where solution of unperturbed part of reduced equation will be in the form

$$u = e^{\left(-\frac{bx^2}{a^2}\right)} erf\left(-\frac{\sqrt{bc_1}x}{a} + c_2\right).$$

Case 2: For $V_3 = X_3$, using the corresponding characteristic equation and change of variables, we write

$$\frac{dt}{0} = \frac{dx}{e^{-bt}} = \frac{du}{0}, \quad u = g(t),$$
$$u_t = g'(t), \quad u_x = u_{xx} = 0.$$

We reduce the perturbed equation Equation (17) to a first order equation:

$$g'(t) - bg(t) + \varepsilon \left[(\ln t + v)g'(t) + g(t) + \sum_{k=1}^{\infty} \frac{(-t)^k}{k(k+1)!} \frac{\partial^{k+1} g}{\partial t^{k+1}}\right] = 0.$$

$u = c_1 e^{bt}$ is a solution of unperturbed equation $g'(t) - bg(t) = 0$.

Case 3: For $V_5 = X_2 + X_3$, the reduced equation is:

$$g' - \frac{1}{2}a^2 e^{2bt} g - bg - \varepsilon \left[(\ln t + v)g' + g(1 + bxe^{bt}) + \sum_{k=1}^{\infty} \frac{(-t)^k}{k(k+1)!} \frac{\partial^{k+1}(ge^{xe^{bt}})}{\partial t^{k+1}}\right] = 0.$$

where $u = \exp\left(\frac{a^2 e^{2bt} + 4b^2 t}{4b}\right)$ is a solution of unperturbed equation.

Case 4: For component of one-dimensional optimal system V_4, V_6 and V_7, solutions of unperturbed part of Equation (17) are given in Table 1.

Table 1. Solutions for unperturbed part of equation Equation (17).

V_i	u
$V_4 = X_4$	$u = c_1 e^{\frac{-b}{a^2}x^2}$
$V_6 = X_5$	$u = e^{bt}(c_1 + c_2 x e^{bt})$
$V_7 = X_6$	$u = e^{\frac{-b}{a^2}x^2}(c_1 + c_2 x e^{-bt})$

5. Approximate Conservation Laws

We consider approximate nonlinear self-adjointness for a system of perturbed PDEs, see, e.g., [24,25] for details. In the rest of this section, we present a formal Lagrange of perturbed Equation (12) and obtain conservation laws.

5.1. Basic Definitions for Constructing Conservation Laws

Let $\mathcal{L}$ be the formal Lagrange of Equation (12):

$$\mathcal{L} \approx \mathcal{L}_{(0)} + \varepsilon \mathcal{L}_{(1)} \equiv v P_{(0)} + \varepsilon v P_{(1)}, \tag{22}$$

hence, the adjoint equations of Equation (12) are defined as

$$\begin{aligned}\frac{\delta \mathcal{L}}{\delta u} &= P^*_{(0)}(x, u, v, u_{(1)}, v_{(1)}, \ldots) \\ &+ \varepsilon P^*_{(1)}(x, u, v, \ldots, D^{c+1}_{x^1} u, D^{c+1}_{x^1} v, D^{c+2}_{x^1} u, D^{c+3}_{x^1} v, \ldots) \approx 0,\end{aligned} \tag{23}$$

where v_i represents all i^{th}-order derivatives of variable v with respect to x, $\frac{\delta}{\delta u}$ is the variational derivative written in terms of the total derivative operator D_i:

$$\frac{\delta}{\delta u} = \frac{\partial}{\partial u} + \sum_{s=1}^{\infty}(-1)^s D_{i_1} \ldots D_{i_s} \frac{\partial}{\partial u_{i_1 \ldots i_s}}.$$

D_i indicates the operator of total differentiation with respect to x^i:

$$D_i = \frac{\partial}{\partial x^i} + u_i \frac{\partial}{\partial u} + v_i \frac{\partial}{\partial v} + \sum_{s=1}^{\infty} \left[u_{ii_1 \ldots i_s} \frac{\partial}{\partial u_{i_1 \ldots i_s}} + v_{ii_1 \ldots i_s} \frac{\partial}{\partial v_{i_1 \ldots i_s}} \right].$$

If we consider

$$v \approx \varphi_{(0)}(x, u) + \varepsilon \varphi_{(1)}(x, u) \neq 0, \tag{24}$$

we have

$$\mathcal{L} \approx \varphi_{(0)} P_{(0)} + \varepsilon \left(\varphi_{(1)} P_{(0)} + \varphi_{(0)} P_{(1)} \right),$$

and if it satisfies the nonlinear self adjoint condition:

$$P^*_0|_{v \approx \varphi_{(0)} + \varepsilon \varphi_{(1)}} + \varepsilon P^*_{(1)}|_{v \approx \varphi_{(0)}} \approx \gamma_{(0)} P_{(0)} + \varepsilon \left(\gamma_{(1)} P_{(0)} + \gamma_{(0)} P_{(1)} \right). \tag{25}$$

In which $\gamma_{(0)}$ and $\gamma_{(1)}$ are to be determined coefficients.

Any approximate symmetry Equation (13) of Equation (12) leads to a conservation law

$$D_i(C^i) = 0, \quad C^i \approx C^i_{(0)} + \varepsilon C^i_{(1)},$$

where the components C^i are obtained by

$$
\begin{aligned}
C^i_{(0)} &= W_{(0)}\left(\frac{\partial \mathcal{L}_{(0)}}{\partial u_i} + \sum_{s=1}^{c-1}(-1)^s D_{i_1}\ldots D_{i_s}\frac{\partial \mathcal{L}_{(0)}}{\partial u_{ii_1\ldots i_s}}\right)\\
&+ \sum_{r=1}^{c-1} D_{k_1}\ldots D_{k_r}\left(W_{(0)}\right)\left[\frac{\partial \mathcal{L}_{(0)}}{\partial u_{ik_1\ldots k_r}} + \sum_{s=1}^{c-r-1}(-1)^s D_{i_1}\ldots D_{i_s}\frac{\partial \mathcal{L}_{(0)}}{\partial u_{ik_1\ldots k_r i_1\ldots i_s}}\right],
\end{aligned} \tag{26}
$$

$$
\begin{aligned}
C^i_{(1)} &= W_{(1)}\left(\frac{\partial \mathcal{L}_{(0)}}{\partial u_i} + \sum_{s=1}^{c-1}(-1)^s D_{i_1}\ldots D_{i_s}\frac{\partial \mathcal{L}_{(0)}}{\partial u_{ii_1\ldots i_s}}\right)\\
&+ \sum_{r=1}^{c-1} D_{k_1}\ldots D_{k_r}\left(W_{(1)}\right)\left[\frac{\partial \mathcal{L}_{(0)}}{\partial u_{ik_1\ldots k_r}} + \sum_{s=1}^{c-r-1}(-1)^s D_{i_1}\ldots D_{i_s}\frac{\partial \mathcal{L}_{(0)}}{\partial u_{ik_1\ldots k_r i_1\ldots i_s}}\right]\\
&+ W_{(0)}\left(\frac{\partial \mathcal{L}_{(1)}}{\partial u_i} + \sum_{s=1}^{\infty}(-1)^s D_{i_1}\ldots D_{i_s}\frac{\partial \mathcal{L}_{(1)}}{\partial u_{ii_1\ldots i_s}}\right)\\
&+ \sum_{r=1}^{\infty} D_{k_1}\ldots D_{k_r}\left(W_{(0)}\right)\left[\frac{\partial \mathcal{L}_{(1)}}{\partial u_{ik_1\ldots k_r}} + \sum_{s=1}^{\infty}(-1)^s D_{i_1}\ldots D_{i_s}\frac{\partial \mathcal{L}_{(1)}}{\partial u_{ik_1\ldots k_r i_1\ldots i_s}}\right].
\end{aligned} \tag{27}
$$

In which $W_{(0)} = \theta_{(0)} - \zeta^i_{(0)} u_i$, $W_{(1)} = \theta_{(1)} - \zeta^i_{(1)} u_i$.

5.2. Approximate Conservation Laws for pfPE

By choosing approximate formal Lagrange

$$
\begin{aligned}
\mathcal{L} \quad & v(x,t,u)\left(P_{(0)} + \varepsilon P_{(1)}\right) =\approx v(x,t,u)\left[u_t - \frac{1}{2}a^2 u_{xx} - bu - bxu_x\right.\\
&\left. + \varepsilon\left((\ln t + v)u_t + \frac{u}{t} + \sum_{k=1}^{\infty}\frac{(-t)^k}{k(k+1)!}u_t^{(k+1)}\right)\right],
\end{aligned} \tag{28}
$$

where

$$
v = \varphi_0(x,t,u) + \varepsilon \varphi_1(x,t,u), \tag{29}
$$

we obtain adjoint equation using Equation (23) as:

$$
P^* \approx -v_t + bxv_x - \frac{1}{2}a^2 v_{xx} - \varepsilon\left[v_t(\ln t + v) + \sum_{k=1}^{\infty}\frac{(-t)^k}{k(k+1)!}D_t^{(k+1)}(vt^k)\right]. \tag{30}
$$

It is easy to achieve an approximate formal Lagrange by placing Equation (29) into Equation (30), and solving characteristic equation of the Equation (25) with the Maple software, we have

$$
\begin{aligned}
v = (c_1 x e^{bt} + c_2) &+ \varepsilon c_3 x e^{c_1 t}\left(c_4 KummerM(\frac{b-c_1}{2b}, \frac{3}{2}, \frac{bx^2}{a^2})\right.\\
&\left. + c_5 KummerU(\frac{b-c_1}{2b}, \frac{3}{2}, \frac{bx^2}{a^2})\right),
\end{aligned} \tag{31}
$$

and

$$
\mathcal{L} \approx \mathcal{L}_{(0)} + \varepsilon \mathcal{L}_{(1)}, \tag{32}
$$

where

$$\begin{aligned}
\mathcal{L}_{(0)} &= (c_1 x e^{bt} + c_2)(u_t - \frac{1}{2}a^2 u_{xx} - bu - bxu_x), \\
\mathcal{L}_{(1)} &= \varepsilon \Bigg[c_3 x e^{c_1 t} \Big(c_4 \text{Kummer} M(\frac{b-c_1}{2b}, \frac{3}{2}, \frac{bx^2}{a^2}) \\
&\quad + c_5 \text{Kummer} U(\frac{b-c_1}{2b}, \frac{3}{2}, \frac{bx^2}{a^2}) \Big)(u_t - \frac{1}{2}a^2 u_{xx} - bu - bxu_x) \\
&\quad + (c_1 x e^{bt} + c_2) \Big((\ln t + v) u_t + \frac{u}{t} + \sum_{k=1}^{\infty} \frac{(-t)^k}{k(k+1)!} u_t^{(k+1)} \Big) \Bigg].
\end{aligned} \tag{33}$$

Here, $c_1, c_2, c_3, c_4, c_5, a$ and b are arbitrary constants. Applying the formula Equations (26) and (27), we perform all computations to approximate conservation laws. Finally, we obtain

$$\begin{aligned}
C^x_{(0)} &= W_{(0)} \Big(-bx\varphi_{(0)} + \frac{1}{2}a^2 D_x \big(\varphi_{(0)} \big) \Big) - \frac{1}{2}a^2 \varphi_{(0)} D_x(W_{(0)}), \\
C^t_{(0)} &= W_{(0)} \varphi_{(0)}, \\
C^x_{(1)} &= W_{(1)} \Big(-bx\varphi_{(0)} + \frac{1}{2}a^2 D_x \big(\varphi_{(0)} \big) \Big) - \frac{1}{2}a^2 \varphi_{(0)} D_x(W_{(1)}) \\
&\quad + W_{(0)} \Big(-bx\varphi_{(1)} + \frac{1}{2}a^2 D_x \big(\varphi_{(1)} \big) \Big) - \frac{1}{2}a^2 \varphi_{(1)} D_x(W_{(0)}),
\end{aligned}$$

$$\begin{aligned}
C^t_{(1)} &= W_{(1)} \varphi_{(0)} + W_{(0)} \Bigg[c_3 x e^{c_1 t} \varphi_{(1)} + \varphi_{(0)} \Big(\ln t + v + \sum_{k=1}^{\infty} \frac{(-t)^k}{k(k+1)!} \Big) \Bigg] \\
&\quad + \varphi_{(0)} \sum_{s=1}^{\infty} (-1)^{(s+1)} D_{st}(W_{(0)}) \Bigg[\sum_{k=s}^{\infty} D_{(k-s)t} \frac{t^k}{k(k+1)!} \Bigg].
\end{aligned}$$

where

$$C^x = C^x_{(0)} + \varepsilon C^x_{(1)}, \qquad C^t = C^t_{(0)} + \varepsilon C^t_{(1)}.$$

1. For $X_1 = \partial_t$, we have $W_{(0)} = -u_t$, $W_{(1)} = 0$, the components of approximate conservation laws are:

$$\begin{aligned}
C^x &= u_t \Big(bx\varphi_{(0)} - \frac{1}{2}a^2 c_1 e^{bt} \Big) + \frac{1}{2}a^2 u_{xt} \varphi_{(0)} \\
&\quad + \varepsilon \Bigg[u_t \Big(bx\varphi_{(1)} - \frac{1}{2}a^2 D_x \varphi_{(1)} \Big) + \frac{1}{2}a^2 \varphi_{(1)} u_{xt} \Bigg],
\end{aligned}$$

$$\begin{aligned}
C^t &= - u_t \varphi_{(0)} + \varepsilon \Bigg[- u_t \varphi_{(1)} + (\ln t + v) \varphi_{(0)} \sum_{k=1}^{\infty} D_{kt} \frac{(-t)^k}{k(k+1)!} \\
&\quad - \varphi_{(0)} \sum_{s=1}^{\infty} D_{st}(u_t) \Big(\sum_{k=s}^{\infty} D_{(k-s)t} \frac{t^k}{k(k+1)!} \Big) \Bigg].
\end{aligned}$$

2. For $X_2 = u \partial_u$, $W_{(0)} = u$ and $W_{(1)} = 0$, we have:

$$\begin{aligned}
C^x &= + u \Big(-bx\varphi_{(0)} + \frac{1}{2}c_1 a^2 e^{bt} \Big) - \frac{1}{2}a^2 u_x \varphi_{(0)} \\
&\quad + \varepsilon \Bigg[u \Big(-bx\varphi_{(1)} + \frac{1}{2}a^2 D_x \varphi_{(1)} \Big) - \frac{1}{2}a^2 u_x \varphi_{(1)} \Bigg],
\end{aligned}$$

$$C^t = + u\varphi_{(0)} + \varepsilon\left[u\left(\varphi_{(1)} + \varphi_{(0)}(\ln t + \nu + \sum_{k=1}^{\infty}D_{kt}\frac{(-t)^k}{k(k+1)!})\right)\right.$$
$$\left. + \varphi_{(0)}\sum_{s=1}^{\infty}(-1)^{s+1}D_{st}(u)\left(\sum_{k=s}^{\infty}D_{(k-s)t}\frac{t^k}{k(k+1)!}\right)\right].$$

3. For $X_3 = e^{-bt}\partial_x$, $W_{(0)} = -e^{-bt}u_x$ and $W_{(1)} = 0$, we have:

$$C^x = + e^{-bt}u_x(bx\varphi_{(0)} - \frac{1}{2}c_1a^2e^{bt}) + \frac{1}{2}a^2e^{-bt}u_{xx}\varphi_{(0)}$$
$$+ e^{-bt}u_x(bx\varphi_{(1)} - \frac{1}{2}a^2D_x\varphi_{(1)}) + \frac{1}{2}a^2e^{-bt}u_{xx}\varphi_{(1)},$$

$$C^t = - u_x e^{-bt}\varphi_{(0)} - \varepsilon\left[u_x e^{-bt}\left(\varphi_{(1)} + \varphi_{(0)}(\ln t + \nu + \sum_{k=1}^{\infty}\frac{(-t)^k}{k(k+1)!})\right)\right.$$
$$\left. + \varphi_{(0)}\sum_{s=1}^{\infty}(-1)^{s+1}D_{st}(u_x e^{-bt})\left(\sum_{k=s}^{\infty}D_{(k-s)t}\frac{t^k}{k(k+1)!}\right)\right].$$

4. For $X_4 = e^{bt}\partial_x - \frac{2b}{a^2}xue^{bt}\partial_u$, $W_{(0)} = -\frac{2b}{a^2}xue^{bt} - e^{bt}u_x$ and $W_{(1)} = 0$, therefore:

$$C^x = e^{bt}\left[(\frac{2b}{a^2}xu + u_x)(bx\varphi_{(0)} - \frac{1}{2}c_1a^2e^{bt}) + (\frac{2b}{a^2}u + u_{xx})(\frac{1}{2}a^2\varphi_{(0)})\right.$$
$$\left. + \varepsilon(\frac{2b}{a^2}xu + u_x)(bx\varphi_{(1)} - \frac{1}{2}a^2D_x\varphi_{(1)}) + (\frac{2b}{a^2}u + u_{xx})(\frac{1}{2}a^2\varphi_{(1)})\right],$$

$$C^t = - e^{bt}\left[(\frac{2b}{a^2}xu + u_x)\varphi_{(0)}\right.$$
$$+ \varepsilon\left((\frac{2b}{a^2}xu + u_x)\left(\varphi_{(1)} + \varphi_{(0)}(\ln t + \nu + \sum_{k=1}^{\infty}D_{kt}\frac{(-t)^k}{k(k+1)!})\right)\right.$$
$$\left.\left. + \varphi_{(0)}\sum_{s=1}^{\infty}(-1)^{s+1}D_{st}(\frac{2b}{a^2}xue^{bt} - e^{bt}u_x)\left(\sum_{k=s}^{\infty}D_{(k-s)t}\frac{t^k}{k(k+1)!}\right)\right)\right].$$

5. For $X_5 = e^{-2bt}(\partial_t - bx\partial_x + bu\partial_u)$, $W_{(0)} = e^{-2bt}(bu - bxu_x - u_t)$ and $W_{(1)} = 0$, so we have:

$$C^x = -e^{-2bt}\left[(bu - bxu_x - u_t)(bx\varphi_{(0)} + \frac{1}{2}c_1a^2e^{bt}) - \frac{1}{2}a^2\varphi_{(0)}(bxu_{xx} + u_{xt})\right.$$
$$\left. + \varepsilon\left((bu - bxu_x - u_t)(bx\varphi_{(1)} - \frac{1}{2}a^2D_x\varphi_{(1)}) - \frac{1}{2}a^2\varphi_{(1)}(bxu_{xx}+u_{xt})\right)\right],$$

$$C^t = e^{-2bt}\varphi_{(0)}(bu - bxu_x - u_t)$$
$$+ \varepsilon\left[e^{-2bt}(bu - bxu_x - u_t)\left(\varphi_{(1)} + \varphi_{(0)}(\ln t + \nu)(\sum_{k=1}^{\infty}D_{kt}\frac{(-t)^k}{k(k+1)!})\right)\right.$$
$$\left. + \varphi_{(0)}\sum_{s=1}^{\infty}(-1)^{s+1}D_{st}(e^{-2bt}(bu - bxu_x - u_t))\left(\sum_{k=s}^{\infty}D_{(k-s)t}\frac{t^k}{k(k+1)!}\right)\right].$$

6. For $X_6 = e^{2bt}(\partial_t + bx\partial_x - \frac{2b^2}{a^2}x^2 u\partial_u)$, $W_{(0)} = -e^{2bt}(\frac{2b^2}{a^2}x^2 u + u_t + bxu_x)$ and $W_{(1)} = 0$, we have:

$$\begin{aligned}
C^x &= e^{2bt}\left[(\frac{2b^2}{a^2}x^2 u + u_t + bxu_x)(\frac{1}{2}c_1 a^2 e^{bt} - bx\varphi_{(0)})\right. \\
&+ \frac{1}{2}a^2\varphi_{(0)}\left(\frac{4b^2}{a^2}xu + \frac{2b^2 x^2}{a^2}u_x + bu_x + bxu_{xx} + u_{xt}\right) \\
&+ \varepsilon\left((\frac{2b^2}{a^2}x^2 u + u_t + bxu_x)(bx\varphi_{(1)} - \frac{1}{2}a^2 D_x\varphi_{(1)})\right. \\
&\left.\left.+ \frac{1}{2}a^2\varphi_{(1)}\left(\frac{4b^2}{a^2}xu + \frac{2b^2 x^2}{a^2}u_x + bu_x + bxu_{xx} + u_{xt}\right)\right)\right], \\
C^t &= -e^{2bt}(\frac{2b^2}{a^2}x^2 u + u_t + bxu_x)\varphi_{(0)} \\
&- \varepsilon\left[e^{2bt}(\frac{2b^2}{a^2}x^2 u + u_t + bxu_x)\left(\varphi_{(1)} + \varphi_{(0)}(\ln t + \nu)(\sum_{k=1}^{\infty} D_{kt}\frac{(-t)^k}{k(k+1)!})\right)\right. \\
&\left.+ \varphi_{(0)}\sum_{s=1}^{\infty}(-1)^{s+1}D_{st}(e^{2bt}(\frac{2b^2}{a^2}x^2 u + u_t + bxu_x))\left(\sum_{k=s}^{\infty} D_{(k-s)t}\frac{t^k}{k(k+1)!}\right)\right].
\end{aligned}$$

7. For $Y_1 = \frac{1}{a^2}\varepsilon\partial_t$, $W_{(0)} = 0$ and $W_{(1)} = \frac{-1}{a^2}\varepsilon u_t$, we have:

$$\begin{aligned}
C^x &= \varepsilon\left[\frac{1}{a^2}u_t(bx\varphi_{(0)} - \frac{1}{2}c_1 a^2 e^{bt}) + \frac{1}{2}u_{xt}\varphi_{(0)}\right], \\
C^t &= -\frac{1}{a^2}\varepsilon u_t \varphi_{(0)}.
\end{aligned}$$

8. For $Y_2 = \varepsilon u\partial_u$, $W_{(0)} = 0$ and $W_{(1)} = \varepsilon u$, we have:

$$\begin{aligned}
C^{(x)} &= \varepsilon\left[u(\frac{1}{2}c_1 a^2 e^{bt} - bx\varphi_{(0)}) - \frac{1}{2}a^2 u_x \varphi_{(0)}\right], \\
C^{(t)} &= \varepsilon u\varphi_{(0)}.
\end{aligned}$$

9. For $Y_3 = \frac{-1}{a^2}\varepsilon e^{-bt}\partial_x$, $W_{(0)} = 0$ and $W_{(1)} = \frac{1}{a^2}\varepsilon e^{-bt}u_x$, we have:

$$\begin{aligned}
C^x &= -\varepsilon\left[\frac{1}{a^2}e^{-bt}u_x\left(bx\varphi_{(0)} - \frac{1}{2}c_1 a^2 e^{bt}\right) + \frac{1}{2}e^{-bt}u_{xx}\varphi_{(0)}\right], \\
C^t &= \frac{1}{a^2}\varepsilon e^{-bt}u_x\varphi_{(0)}.
\end{aligned}$$

10. For $Y_4 = \frac{-1}{a^2}\varepsilon e^{bt}(\partial_x + 2bxu\partial_u)$, $W_{(0)} = 0$ and $W_{(1)} = \frac{1}{a^2}\varepsilon e^{bt}(u_x - 2bxu)$, we have:

$$C^x = \frac{1}{a^2}\varepsilon e^{bt}\left[(u_x - 2bxu)\left(\frac{1}{2}c_1 a^2 e^{bt} - bx\varphi_{(0)}\right) + \frac{a^2}{2}\varphi_{(0)}(2bu + 2bxu_x - u_{xx})\right],$$

$$C^t = \frac{1}{a^2}\varepsilon e^{bt}(u_x - 2bxu)\varphi_{(0)}.$$

11. For $Y_5 = \frac{1}{a^2}\varepsilon e^{-2bt}(\partial_t + bx\partial_x + bu\partial_u)$, $W_{(0)} = 0$ and
    $W_{(1)} = \frac{1}{a^2}\varepsilon e^{-2bt}(bu - u_t - bxu_x)$, we have:

$$
\begin{aligned}
C^x &= \frac{1}{a^2}\varepsilon e^{-2bt}\left[(bu - u_t - bxu_x)\left(\frac{1}{2}c_1 a^2 e^{bt} - bx\varphi_{(0)}\right)\right.\\
&\quad + \left.\frac{1}{2}a^2\varphi_{(0)}(u_{xt} + bxu_{xx})\right],
\end{aligned}
$$

$$
C^t = \frac{1}{a^2}\varepsilon e^{-2bt}(bu - u_t - bxu_x)\varphi_{(0)}.
$$

12. For $Y_6 = \frac{1}{a^2}\varepsilon e^{2bt}(\partial_t - bx\partial_x - 2b^2 x^2 u\partial_u)$, $W_{(0)} = 0$ and
    $W_{(1)} = \frac{1}{a^2}\varepsilon e^{2bt}(bxu_x - u_t - 2b^2 x^2 u)$, we have:

$$
\begin{aligned}
C^x &= \frac{1}{a^2}\varepsilon e^{2bt}(bxu_x - u_t - 2b^2 x^2 u)\left(\frac{1}{2}c_1 a^2 e^{bt} - bx\varphi_{(0)}\right)\\
&\quad + \frac{1}{a^2}\varphi_{(0)}(4b^2 xu + 2b^2 x^2 u_x + u_{xt} - bu_x - bxu_{xx}),\\
C^t &= \frac{1}{a^2}\varepsilon e^{2bt}(bxu_x - u_t - 2b^2 x^2 u)\varphi_{(0)}.
\end{aligned}
$$

6. Conclusions and Outlook

We presented a new approach for calculating new exact analytical solutions of parameter containing fractional-order equations. Using the nonlinear self-adjoint notion, approximate solutions, conservation laws and symmetries for these equations are obtained. Computational results indicate the strength of new method. We will apply the method to fractional-stochastic differential equations in a future work.

Author Contributions: Formal analysis, E.L.; Funding acquisition, Y.-M.C.; Investigation, N.K.; Supervision, M.I.; Writing—original draft, M.A.A. All authors have read and agreed to the published version of the manuscript.

Funding: The work was supported by the Natural Science Foundation of China (Grant Nos. 61673169, 11301127, 11701176, 11626101, 11601485).

Conflicts of Interest: The authors declare that they have no known competing financial interests or personal relationships that could have appeared to influence the work reported in this paper.

References

1. Samko, S.G.; Ross, B. Integration and differentiation to a variable fractional order. *Integral Transform. Spec. Funct.* **1993**, *1*, 277–300. [CrossRef]
2. Kiryakova, V.S. *Generalized Fractional Calculus and Applications*; CRC Press: Boca Raton, FL, USA 1993.
3. Lashkarian, E.; Hejazi, S.R.; Dastranj, E. Conservation laws of (3+α)-dimensional time-fractional diffusion equation. *Comput. Math. Appl.* **2018**, *75*, 740–754. [CrossRef]
4. Lashkarian, E.; Hejazi, S.R. Group analysis of the time fractional generalized diffusion equation. *Phys. A Stat. Its Appl.* **2017**, *479*, 572–579. [CrossRef]
5. Jafari, H.; Kadkhoda, N.; Baleanu, D. Fractional Lie group method of the time-fractional Boussinesq equation. *Nonlinear Dyn.* **2015**, *81*, 1569–1574. [CrossRef]
6. Kilbas, A.A.; Srivastava, H.M.; Trujillo, J.J. *Theory and Applications of Fractional Differential Equations*; Elsevier Science Limited: New York, NY, USA 2006; Volume 204.
7. Podlubny, I. *Fractional Differential Equations: An Introduction to Fractional Derivatives, Fractional Differential Equations to Methods of Their Solution and Some of Their Applications*; Elsevier: New York, NY, USA, 1998; Volume 198.
8. Fuente, D.; Romero, A. Uniformly accelerated motion in General Relativity: Completeness of inextensible trajectories. *Gen. Relativ. Gravit.* **2015**, *47*, 33. [CrossRef]

9. Euler, M.; Euler, N.; Kohler, A. On the construction of approximate solutions for a multidimensional nonlinear heat equation. *J. Phys. A Math. Gen.* **1994**, *27*, 2083. [CrossRef]
10. Mahomed, F.M.; Qu, C. Approximate conditional symmetries for partial differential equations. *J. Phys. A Math. Gen.* **2000**, *3*, 343. [CrossRef]
11. Ibragimov, N.H. Nonlinear self-adjointness in constructing conservation laws. *arXiv* **2011**, arXiv:1109.1728.
12. Johnpillai, A.G.; Kara, A.H.; Mahomed, F.M. A basis of approximate conservation laws for PDEs with a small parameter. *Int. J. Non-Linear Mech.* **2006**, *41*, 830–837. [CrossRef]
13. Johnpillai, A.G.; Kara, A.H.; Mahomed, F.M. Approximate Noether-type symmetries and conservation laws via partial Lagranges for PDEs with a small parameter. *J. Comput. Appl. Math.* **2009**, *223*, 508–518. [CrossRef]
14. Baikov, V.A.; Gazizov, R.K.; Ibragimov, N.K. Perturbation methods in group analysis. *J. Sov. Math.* **1991**, *55*, 1450–1490. [CrossRef]
15. Fushchich, W.I.; Shtelen, W.M. On approximate symmetry and approximate solutions of the nonlinear wave equation with a small parameter. *J. Phys. A Math. Gen.* **1989**, *22*, L887. [CrossRef]
16. Lukashchuk, S.Y. Constructing conservation laws for fractional-order integro-differential equations. *Theor. Math. Phys.* **2015**, *184*, 1049–1066. [CrossRef]
17. Johnpillai, A.G.; Kara, A.H. Variational formulation of approximate symmetries and conservation laws. *Int. J. Theor. Phys.* **2001**, *40*, 1501–1509. [CrossRef]
18. Burde, G.I. On the use of the lie group technique for differential equations with a small parameter: Approximate solutions and integrable equations. *Phys. At. Nucl.* **2002**, *65*, 990–995. [CrossRef]
19. Reza Hejazi, S.; Hosseinpour, S.; Lashkarian, E. Approximate symmetries, conservation laws and numerical solutions for a class of perturbed linear wave type system. *Quaest. Math.* **2019**, *42*, 1–17. [CrossRef]
20. Lukashchuk, S.Y. Conservation laws for time-fractional subdiffusion and diffusion-wave equations. *Nonlinear Dyn.* **2015**, *80*, 791–802. [CrossRef]
21. Montagnon, C. A closed solution to the Fokker Planck equation applied to forecasting. *Phys. A Stat. Mech. Its Appl.* **2015**, *420*, 14–22. [CrossRef]
22. Smirnov, A.P.; Shmelev, A.B.; Sheinin, E.Y. Analysis of Fokker Planck approach for foreign exchange market statistics study. *Phys. A Stat. Mech. Its Appl.* **2004**, *344*, 203–206. [CrossRef]
23. Olver, P.J. *Applications of Lie Groups to Differential Equations*; Springer Science and Business Media: Berlin, Germany 2012; Volume 107.
24. Ibragimov, N.H. Nonlinear self-adjointness and conservation laws. *J. Phys. A Math. Theor.* **2011**, *44*, 432002. [CrossRef]
25. Ibragimov, N.K.; Avdonina, E.D. Nonlinear self-adjointness, conservation laws, and the construction of solutions of partial differential equations using conservation laws. *Russ. Math. Surv.* **2013**, *68*, 889. [CrossRef]

© 2020 by the authors. Licensee MDPI, Basel, Switzerland. This article is an open access article distributed under the terms and conditions of the Creative Commons Attribution (CC BY) license (http://creativecommons.org/licenses/by/4.0/).

Article

New Soliton Solutions of Fractional Jaulent-Miodek System with Symmetry Analysis

Subhadarshan Sahoo [1], Santanu Saha Ray [2], Mohamed Aly Mohamed Abdou [3,4], Mustafa Inc [5,6,*] and Yu-Ming Chu [7,8,*]

[1] Department of Mathematics, Kalinga Institute of Industrial Technology, Bhubaneswar-751024, India; subhadarshan.sahoofma@kiit.ac.in or subha.bapi25@gmail.com
[2] Department of Mathematics, National Institute of Technology, Rourkela-769008, India; saharays@nitrkl.ac.in or santanusaharay@yahoo.com
[3] Department of Physics, College of Sciences, University of Bisha, P.O. Box 344, Bisha 61922, Saudi Arabia; m_abdou@mans.edu.eg or m_abdou_eg@yahoo.com
[4] Department of Physics, Theoretical Research Group, Science Faculty, Mansoura University, Mansoura 35516, Egypt
[5] Department of Mathematics, Science Faculty, Firat University, Elazig 23119, Turkey
[6] Department of Medical Research, China Medical University Hospital, China Medical University, Taichung 411, Taiwan
[7] Department of Mathematics, Huzhou University, Huzhou 313000, China
[8] Hunan Provincial Key Laboratory of Mathematical Modeling and Analysis in Engineering, Changsha University of Science & Technology, Changsha 410114, China
* Correspondence: minc@firat.edu.tr (M.I.); chuyuming@zjhu.edu.cn or chuyuming2005@126.com (Y.-M.C.)

Received: 12 May 2020; Accepted: 5 June 2020; Published: 12 June 2020

Abstract: New soliton solutions of fractional Jaulent-Miodek (JM) system are presented via symmetry analysis and fractional logistic function methods. Fractional Lie symmetry analysis is unified with symmetry analysis method. Conservation laws of the system are used to obtain new conserved vectors. Numerical simulations of the JM equations and efficiency of the methods are presented. These solutions might be imperative and significant for the explanation of some practical physical phenomena. The results show that present methods are powerful, competitive, reliable, and easy to implement for the nonlinear fractional differential equations.

Keywords: fractional Jaulent-Miodek (JM) system; fractional logistic function method; symmetry analysis

1. Introduction

Integral and derivative operators of any arbitrary order are the basis of fractional calculus, which has been of great interest for researchers due to its dynamic behavior and exact description of nonlinear complex phenomena in numerous fields in science and engineering [1–6]. Analytical methods have played an essential role for Fractional partial differential equations (FPDEs) [1–4]. Lie symmetry analysis also gives a powerful and effectual implement for generating invariant solutions. The theory of symmetry analysis is based on the invariance of variables [7–14]. Hence, the study of symmetry analysis has been made a huge interest for researchers during past decades.

Time-fractional coupled Jaulent-Miodek (JM) type equations [15–17] is considered as:

$$D_t^\alpha u + u_{xxx} + \frac{3}{2}vv_{xxx} + \frac{9}{2}v_x v_{xx} - 6uu_x - 6uvv_x - \frac{3}{2}u_x v^2 = 0 \qquad (1)$$

and

$$D_t^\alpha v + v_{xxx} - 6u_x v - 6uv_x - \frac{15}{2}v_x v^2 = 0 \qquad (2)$$

where $0 < \alpha \leq 1$ denotes the fractional-order derivative.

The coupled JM equations were first introduced by Jaulent and Miodek [18] by using inverse scattering transform with the help of energy dependent Schrödinger potentials. The Equations (1) and (2) also have a relation with Euler-Darboux equation, which has been presented by Matsuno [19]. The Darboux transformation of the JM spectral problem has been studied by Xu [20]. By using hereditary symmetries, Ruan and Lou [21] have presented the symmetries of Jaulent-Miodek hierarchy. The sech and tanh–coth methods have been used by Wazwaz [22] and some more methods like homotopy analysis [23], exp-function [24], extended tanh [25], hyperbolic tangent [26] were presented in the literature for approximate and exact solutions of classical coupled Jaulent-Miodek equation.

A large interest has been focused for the improvement of past methods dealing with solutions of FPDEs. The fractional coupled JM equations play an important role in several areas of science such as fluid mechanics, plasma physics, condense matter physics, optics and associates with energy dependent Schrödinger potential [27–32]. As the practical application of fractional Jaulent–Miodek (JM) system, the Wang and Xia has studied its super-Hamiltonian structure using fractional supertrace identity [33].

Some of these methods for solving fractional coupled JM equation are: method of homotopy perturbation natural transform [34], Sumudu transform [15], residual power series method (RSPM) and q-homotopy analysis method (q-HAM) [17], Hermite wavelet [35], (G'/G)-expansion and hyperbolic tangent [16].

This article deals with fractional coupled JM system by utilizing an original fractional logistic function method [36], which has been presented in Section 3. Moreover, in the corresponding section, the numerical simulation has been done for analyzing the physical properties of the solutions. In Section 4, the symmetry analysis with conservation laws [37,38] for time-fractional coupled JM, equations have been presented. In Section 4, the fractional Lie group analysis method for symmetry properties [39,40] of fractional JM system are applied more precisely. Furthermore, conservation laws [37,41] also have been presented in order to get a new conserved vector by utilizing theorems of conservation law.

2. Theory of Fractional Operators

2.1. Riemann–Liouville (RL) Fractional Derivative

The fractional order Riemann–Liouville (RL) derivative of order $\alpha (>0)$ is defined as [1,3]

$$D_t^\alpha f(t) = \begin{cases} \frac{1}{\Gamma(m-\alpha)} \frac{d^m}{dt^m} \int_0^t (t-\tau)^{(m-\alpha-1)} f(\tau) d\tau & \text{if } m-1 < \alpha < m, \ m \in \mathbb{N}, \\ \frac{d^m f(t)}{dt^m} & \text{if } \alpha = m, \ m \in \mathbb{N}, \end{cases} \quad (3)$$

Riemann–Liouville (RL) derivative of order α (>0) has subsequent property [1–3] is given as:

$$D^\alpha t^\beta = \frac{\Gamma(\beta+1)}{\Gamma(\beta-\alpha+1)} t^{\beta-\alpha}, \quad \beta > \alpha - 1. \quad (4)$$

2.2. Local Fractional-Order Derivative

Assume $h(\overleftarrow{x}) \in C_\alpha(m,n)$, where $C_\alpha(m,n)$ denotes α times differentiable with each derivative continuous in (m,n). Then, the derivative with fractional order α at $\overleftarrow{x} = \overleftarrow{x}_0$ is defined as [42,43]

$$h^{(\alpha)}(\overleftarrow{x}_0) = \left. \frac{d^\alpha h(\overleftarrow{x})}{d\overleftarrow{x}^\alpha} \right|_{\overleftarrow{x}=\overleftarrow{x}_0} = \lim_{\overleftarrow{x} \to \overleftarrow{x}_0} \frac{\Delta^\alpha(h(\overleftarrow{x}) - h(\overleftarrow{x}_0))}{(\overleftarrow{x} - \overleftarrow{x}_0)^\alpha} \quad (5)$$

where $\Delta^\alpha(h(\overleftarrow{x}) - h(\overleftarrow{x}_0)) \cong \Gamma(1+\alpha)(h(\overleftarrow{x}) - h(\overleftarrow{x}_0))$ and $0 < \alpha \leq 1$.

And has following property [42,43]:

If $z(\overleftarrow{x}) = (h \circ u)(\overleftarrow{x})$, where $u(\overleftarrow{x}) = f(\overleftarrow{x})$, then

$$\frac{d^\alpha z(\overleftarrow{x})}{dx^{-\alpha}} = h^{(1)}\big(f(\overleftarrow{x})\big) f^{(\alpha)}(\overleftarrow{x}) \tag{6}$$

when $h^{(1)}\big(f(\overleftarrow{x})\big)$ and $f^{(\alpha)}(\overleftarrow{x})$ exist.

3. The Brief Descriptions of the Fractional Logistic Function Method and Implementations

3.1. Brief Description of the Proposed Method

The section emphasizes describing a comparatively new analytic method for getting solutions for the FPDEs. The procedure for the proposed method has been described in the following manner:

Step 1:

The FPDE is given as:

$$Q(u, D_t^\alpha u, \ldots, u_x, u_{xx}, u_{xxx}, \ldots) = 0, \ 0 < \alpha \leq 1, \tag{7}$$

where $u(x,t)$ is a function.

Step 2:

Solution of Equation (7) is presented as

$$u(x,t) = U(\xi), \ \xi = kx - \frac{\gamma \, t^\alpha}{\Gamma(\alpha+1)}, \tag{8}$$

where γ and k are parameters.

Then, (6) [44,45] can reduce the fractional derivative into the following form

$$D_t^\alpha u = \sigma_t U_\xi D_t^\alpha \xi$$

Then, the Equation (7) can be reduced by using Equation (7), by the following form:

$$Q(U, \gamma U', \ldots, kU', k^2 U'', k^3 U''', \ldots) = 0 \tag{9}$$

Step 3:

Here, the exact solution of Equation (7) is mentioned in terms of the polynomial in $\varphi(\xi)$ as follows:

$$U(\xi) = a_0 + \sum_{i=1}^{n} a_i \varphi^i(\xi), \tag{10}$$

where $\varphi(\xi)$ is considered as the sigmoid function or logistic function [46,47], is defined as follows: $\varphi(\xi) = \frac{e^\xi}{1+e^\xi}$ and satisfies the following Riccati equation:

$$\phi_\xi = \phi - \phi^2, \tag{11}$$

and the value of n can be evaluated by using the homogenous balancing principle [48,49]. Moreover, the derivatives of different order for the function $U(\xi)$ can be determined by using Equation (11).

Step 4:

Now, the coefficients a_i are determined by putting Equation (11) into Equation (9) and solving the acquired algebraic equations obtained by equating coefficients of φ^i to 0.

Step 5:

Unknowns obtained in step 4 are written into Equation (10) to get the solutions for Equation (7).

3.2. Soliton Solutions for JM System

The logistic function method is employed for solving Equation (1). By using Equation (8) in Equation (1), we have:

$$-\gamma U'(\xi) + k^3 U'''(\xi) + \frac{3k^3}{2} V(\xi) V'''(\xi) + \frac{9k^3}{2} V'(\xi) V''(\xi) - 6kU(\xi)U'(\xi) - 6kU(\xi)V(\xi)V'(\xi) - \frac{3}{2}kU'(\xi)V^2(\xi) = 0, \tag{12}$$

and

$$-\gamma V'(\xi) + k^3 V'''(\xi) - 6kU'(\xi)V(\xi) - 6kU(\xi)V'(\xi) - \frac{15k}{2} V(\xi) V^2(\xi) = 0, \tag{13}$$

Similar to Equation (10), let us consider the solutions of the governing system are presented by following mathematical equations as

$$U(\xi) = a_0 + \sum_{i=1}^{n} a_i \varphi^i \text{ and } V(\xi) = b_0 + \sum_{i=1}^{m} b_i \varphi^i \tag{14}$$

By means of homogenous balance principle [48,49], we get $n = 2$ and $m = 1$. Thus, the solutions are:

$$U(\xi) = a_0 + a_1 \varphi + a_2 \varphi^2 \text{ and } V(\xi) = b_0 + b_1 \varphi, \tag{15}$$

where φ follows satisfies Equation (11).

Putting Equation (15) with Equation (11) into Equations (12) and (13), equating the obtained coefficient of φ^i to 0, we get:

Set 1:

$$\gamma = \frac{k^3}{4},\ a_0 = -\frac{k^2}{32},\ a_1 = -\frac{3k^2}{8},\ a_2 = \frac{3k^2}{8},\ b_0 = \frac{ik}{2\sqrt{2}},\ b_1 = -\frac{ik}{\sqrt{2}}.$$

For **set 1**, the following hyperbolic solutions can be obtained as

$$\begin{aligned} U_{11} &= -\frac{k^2(\cosh(\xi)+7)}{32(1+\cosh(\xi))} \\ V_{12} &= -\frac{ik\tanh\left(\frac{\xi}{2}\right)}{2\sqrt{2}} \end{aligned} \tag{16}$$

where $\xi = kx - \frac{k^3 \mu^\alpha}{4\Gamma(\alpha+1)}$.

Set 2:

$$\gamma = \frac{k^3}{4},\ a_0 = -\frac{k^2}{32},\ a_1 = -\frac{3k^2}{8},\ a_2 = \frac{3k^2}{8},\ b_0 = -\frac{ik}{\sqrt{2}},\ b_1 = \frac{ik}{\sqrt{2}}$$

For **set 2**, the following hyperbolic solutions can be obtained as

$$\begin{aligned} U_{21} &= -\frac{k^2(\cosh(\xi)+7)}{32(1+\cosh(\xi))} \\ V_{22} &= -\frac{ik(1+3\cosh(\xi)+3\sinh(\xi))}{2\sqrt{2}(1+\cosh(\xi)+\sinh(\xi))} \end{aligned} \tag{17}$$

where $\xi = kx - \frac{k^3 \mu^\alpha}{4\Gamma(\alpha+1)}$.

Set 3:

$$\gamma = \frac{11k^3}{5}, \ a_0 = \frac{k^2}{20}, \ a_1 = -2k^2, \ a_2 = 2k^2, \ b_0 = i\sqrt{5}k, \ b_1 = -2i\sqrt{5}k$$

For **set 3**, the following hyperbolic solutions can be obtained as

$$\begin{aligned} U_{31} &= \frac{k^2(\cosh(\xi)-19)}{20(1+\cosh(\xi))} \\ V_{32} &= -i\sqrt{5}k\tanh\left(\frac{\xi}{2}\right) \end{aligned} \quad (18)$$

where $\xi = kx - \frac{11k^3 t^\alpha}{5\Gamma(\alpha+1)}$.

Set 4:

$$\gamma = \frac{11k^3}{5}, \ a_0 = \frac{k^2}{20}, \ a_1 = -2k^2, \ a_2 = 2k^2, \ b_0 = -i\sqrt{5}k, \ b_1 = 2i\sqrt{5}k$$

For **set 4**, the following hyperbolic solutions can be obtained as

$$\begin{aligned} U_{41} &= \frac{k^2(\cosh(\xi)-19)}{20(1+\cosh(\xi))} \\ V_{42} &= i\sqrt{5}k\tanh\left(\frac{\xi}{2}\right) \end{aligned} \quad (19)$$

where $\xi = kx - \frac{11k^3 t^\alpha}{5\Gamma(\alpha+1)}$.

3.3. Numerical Simulations

This part emphasizes on numerical simulation for the Equations (1) and (2) by the fractional logistic equation method. Furthermore, the Equations (16) and (18) have been used here for generating solutions graphs.

The Figures 1–4 illustrates obtained solutions of governing equations.

Case 1: For $\alpha = 0.1$ (Fractional order)

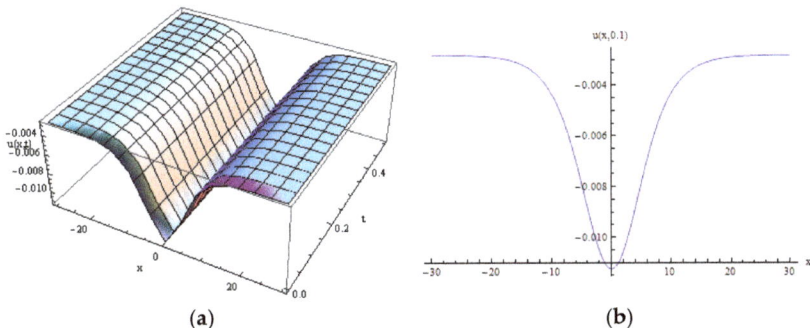

(a) (b)

Figure 1. (a) A three dimensional (3-D) solitary wave figure of $u(x,t)$ in Equation (16) with U_{11}, when $k = 0.3$ and $\alpha = 0.1$, (b) 2-D figure of $u(x,t)$, for $t = 0.1$.

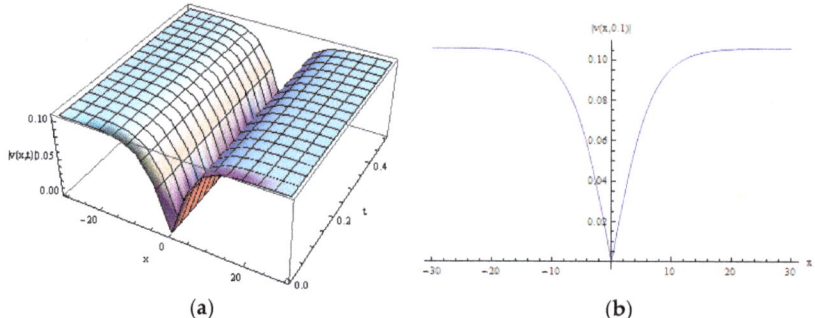

Figure 2. (a) A 3-D solitary wave of $|v(x,t)|$ in Equation (16) with V_{12}, when $k = 0.3$ and $\alpha = 0.1$, (b) 2-D figure of $|v(x,t)|$ for $t = 0.1$.

Case 2: For $\alpha = 0.1$ (Fractional order)

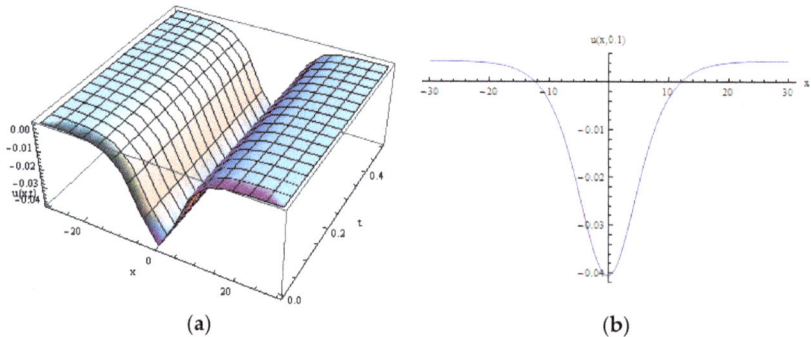

Figure 3. (a) A 3-D solitary wave figure of $u(x,t)$ in Equation (18) as U_{31}, for $k = 0.3$ and $\alpha = 0.1$, (b) 2-D figure of $u(x,t)$ for $t = 0.1$.

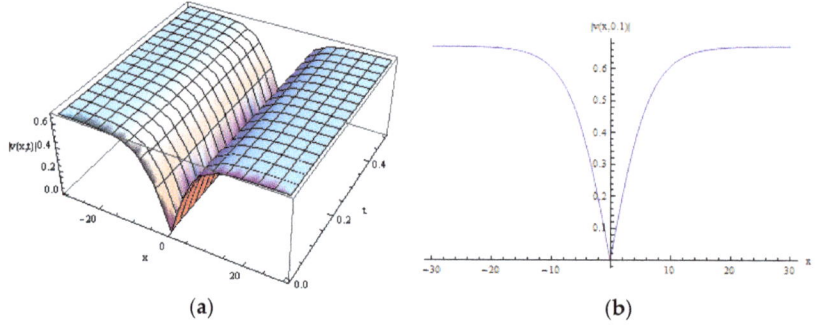

Figure 4. (a) A 3-D solitary wave figure of $|v(x,t)|$ in Equation (16) with V_{32}, for $k = 0.3$ and $\alpha = 0.1$, (b) 2-D figure of $|v(x,t)|$ for $t = 0.1$.

4. Lie Symmetry Analysis Method

4.1. Theory of Symmetry Analysis Method

In this part, the general method for generating the symmetries of FPDEs is discussed by means of fractional Lie symmetry analysis.

Consider

$$D_t^\alpha u = F(t, x, u, u_x, u_{xx}, u_{xxx}, v, v_x, v_{xx}, v_{xxx}, \ldots) \tag{20}$$

$$D_t^\alpha v = G(t, x, u, u_x, u_{xx}, u_{xxx}, v, v_x, v_{xx}, v_{xxx}, \ldots) \tag{21}$$

Let us now consider that the Equations (20) and (21) are invariant in one-parameter Lie group transformation:

$$
\begin{aligned}
\overleftrightarrow{x} &\to x + \varepsilon \xi(t, x, u, v) + O(\varepsilon^2), \\
\overleftrightarrow{t} &\to t + \varepsilon \tau(t, x, u, v) + O(\varepsilon^2), \\
\overleftrightarrow{u} &\to u + \varepsilon \eta(t, x, u, v) + O(\varepsilon^2), \\
\overleftrightarrow{v} &\to v + \varepsilon \vartheta(t, x, u, v) + O(\varepsilon^2), \\
D_t^\alpha \overleftrightarrow{u} &\to D_t^\alpha u + \varepsilon \eta_a^0(t, x, u, v) + O(\varepsilon^2), \\
D_t^\alpha \overleftrightarrow{v} &\to D_t^\alpha v + \varepsilon \vartheta_a^0(t, x, u, v) + O(\varepsilon^2), \\
\frac{\partial \overleftrightarrow{u}}{\partial \overleftrightarrow{x}} &\to \frac{\partial u}{\partial x} + \varepsilon \eta^x(t, x, u, v) + O(\varepsilon^2), \\
\frac{\partial \overleftrightarrow{v}}{\partial \overleftrightarrow{x}} &\to \frac{\partial v}{\partial x} + \varepsilon \vartheta^x(t, x, u, v) + O(\varepsilon^2), \\
\frac{\partial^2 \overleftrightarrow{u}}{\partial \overleftrightarrow{x}^2} &\to \frac{\partial^2 u}{\partial x^2} + \varepsilon \eta^{xx}(t, x, u, v) + O(\varepsilon^2), \\
\frac{\partial^2 \overleftrightarrow{v}}{\partial \overleftrightarrow{x}^2} &\to \frac{\partial^2 v}{\partial x^2} + \varepsilon \vartheta^{xx}(t, x, u, v) + O(\varepsilon^2), \\
\frac{\partial^3 \overleftrightarrow{u}}{\partial \overleftrightarrow{x}^3} &\to \frac{\partial^3 u}{\partial x^3} + \varepsilon \eta^{xxx}(t, x, u, v) + O(\varepsilon^2), \\
\frac{\partial^3 \overleftrightarrow{v}}{\partial \overleftrightarrow{x}^3} &\to \frac{\partial^3 v}{\partial x^3} + \varepsilon \vartheta^{xxx}(t, x, u, v) + O(\varepsilon^2), \\
&\ldots
\end{aligned}
\tag{22}
$$

where $\varepsilon \ll 1$ is considered as a group parameter, $\tau, \eta, \vartheta, \xi$ are infinitesimals. Total expression for η^x, $\eta^{xx}, \eta^{xxx}, \vartheta^x, \vartheta^{xx}$ and ϑ^{xxx} are:

$$
\begin{aligned}
\eta^x &= D_x(\eta) - u_x D_x(\xi) - u_t D_x(\tau), \\
\eta^{xx} &= D_x(\eta^x) - u_{xx} D_x(\xi) - u_{xt} D_x(\tau), \\
\eta^{xxx} &= D_x(\eta^{xx}) - u_{xxx} D_x(\xi) - u_{xxt} D_x(\tau), \\
\vartheta^x &= D_x(\vartheta) - v_x D_x(\xi) - v_t D_x(\tau), \\
\vartheta^{xx} &= D_x(\vartheta^x) - v_{xx} D_x(\xi) - v_{xt} D_x(\tau), \\
\vartheta^{xxx} &= D_x(\vartheta^{xx}) - v_{xxx} D_x(\xi) - v_{xxt} D_x(\tau)
\end{aligned}
\tag{23}
$$

where $D_{x^j} = \frac{\partial}{\partial x^j} + u_j \frac{\partial}{\partial u} + v_j \frac{\partial}{\partial v} + u_{jk} \frac{\partial}{\partial u_k} + v_{jk} \frac{\partial}{\partial u_k} + \ldots$, $j, k = 1, 2, 3, \ldots$ and $u_j = \frac{\partial u}{\partial x^j}, v_j = \frac{\partial v}{\partial x^j}$, $u_{jk} = \frac{\partial^2 u}{\partial x^j \partial x^k}, v_{jk} = \frac{\partial^2 v}{\partial x^j \partial x^k}$ and so on.

$$\mathbf{V} = \xi(t, x, u, v) \frac{\partial}{\partial x} + \tau(t, x, u, v) \frac{\partial}{\partial t} + \eta(t, x, u, v) \frac{\partial}{\partial u} + \vartheta(t, x, u, v) \frac{\partial}{\partial v} \tag{24}$$

V satisfies:

$$\left. \text{Pr}^{(n)} \mathbf{V}(\Delta_1) \right|_{\Delta_1 = 0} = 0 \text{ and } \left. \text{Pr}^{(n)} \mathbf{V}(\Delta_2) \right|_{\Delta_2 = 0} = 0, \, n = 1, 2, \ldots, \tag{25}$$

here, Pr denotes the prolongation for the given vector and

$$\Delta_1 := D_t^\alpha u - F(t, x, u, u_x, u_{xx}, u_{xxx}, v, v_x, v_{xx}, v_{xxx}, \ldots)$$

and

$$\Delta_2 := D_t^\alpha v - G(t, x, u, u_x, u_{xx}, u_{xxx}, v, v_x, v_{xx}, v_{xxx}, \ldots)$$

Now, by considering the usual structure of RL fractional operator, the transformations of system (22) has been formed. We have

$$\tau(x,t,u,v)\big|_{t=0} = 0 \tag{26}$$

By RL derivative, the α-th infinitesimal [50–52] with Equation (26) can be presented as follows:

$$\eta_\alpha^0 = D_t^\alpha(\eta) + \xi D_t^\alpha(u_x) - D_t^\alpha(\xi u_x) + D_t^\alpha(D_t(\tau)u) - D_t^{\alpha+1}(\tau u) + \tau D_t^{\alpha+1}(u)$$

and

$$\vartheta_\alpha^0 = D_t^\alpha(\vartheta) + \xi D_t^\alpha(v_x) - D_t^\alpha(\xi v_x) + D_t^\alpha(D_t(\tau)v) - D_t^{\alpha+1}(\tau v) + \tau D_t^{\alpha+1}(v) \tag{27}$$

where the D_t^α denotes the total fractional differential operator.

We have:

$$D_t^\alpha(f(t)g(t)) = \sum_{m=0}^\infty \binom{\alpha}{m} D_t^{\alpha-m} f(t) D_t^m g(t), \quad \alpha > 0 \tag{28}$$

where

$$\binom{\alpha}{m} = \frac{(-1)^{m-1}\alpha \Gamma(m-\alpha)}{\Gamma(1-\alpha)\Gamma(m+1)}$$

We also have

$$\eta_\alpha^0 = D_t^\alpha(\eta) - \alpha D_t^\alpha(\tau)\frac{\partial^\alpha u}{\partial t^\alpha} - \sum_{n=1}^\infty \binom{\alpha}{n} D_t^n(\xi) D_t^{\alpha-n} u_x - \sum_{n=1}^\infty \binom{\alpha}{n+1} D_t^{n+1}(\tau) D_t^{\alpha-n}(u)$$

and

$$\vartheta_\alpha^0 = D_t^\alpha(\vartheta) - \alpha D_t^\alpha(\tau)\frac{\partial^\alpha v}{\partial t^\alpha} - \sum_{n=1}^\infty \binom{\alpha}{n} D_t^n(\xi) D_t^{\alpha-n} v_x - \sum_{n=1}^\infty \binom{\alpha}{n+1} D_t^{n+1}(\tau) D_t^{\alpha-n}(v) \tag{29}$$

We have:

$$\frac{d^m g(h(t))}{dt^m} = \sum_{k=0}^m \sum_{r=0}^k \binom{k}{r}\frac{1}{k!}[-h(t)]^r \frac{d^m}{dt^m}[h(t)^{k-r}]\frac{d^k g(h)}{dh^k} \tag{30}$$

Now by using Equations (28) and (30) with $f(t) = 1$, we have

$$D_t^\alpha(\eta) = \frac{\partial^\alpha \eta}{\partial t^\alpha} + \eta_u \frac{\partial^\alpha u}{\partial t^\alpha} - u\frac{\partial^\alpha \eta_u}{\partial t^\alpha} + \sum_{n=1}^\infty \binom{\alpha}{n}\frac{\partial^n \eta_u}{\partial t^n} D_t^{\alpha-n}(u) + \mu$$

and

$$D_t^\alpha(\vartheta) = \frac{\partial^\alpha \vartheta}{\partial t^\alpha} + \vartheta_v \frac{\partial^\alpha v}{\partial t^\alpha} - v\frac{\partial^\alpha \eta_v}{\partial t^\alpha} + \sum_{n=1}^\infty \binom{\alpha}{n}\frac{\partial^n \vartheta_v}{\partial t^n} D_t^{\alpha-n}(v) + \lambda \tag{31}$$

where

$$\mu = \sum_{n=2}^\infty \sum_{m=2}^n \sum_{k=2}^m \sum_{r=0}^{k-1} \binom{\alpha}{n}\binom{n}{m}\binom{k}{r}\frac{1}{k!}\frac{t^{n-\alpha}}{\Gamma(n+1-\alpha)}(-u)^r \frac{\partial^m}{\partial t^m}(u^{k-r})\frac{\partial^{n-m+k}\eta}{\partial t^{n-m}\partial u^k}$$

and

$$\lambda = \sum_{n=2}^\infty \sum_{m=2}^n \sum_{k=2}^m \sum_{r=0}^{k-1} \binom{\alpha}{n}\binom{n}{m}\binom{k}{r}\frac{1}{k!}\frac{t^{n-\alpha}}{\Gamma(n+1-\alpha)}(-v)^r \frac{\partial^m}{\partial t^m}(v^{k-r})\frac{\partial^{n-m+k}\vartheta}{\partial t^{n-m}\partial v^k}$$

Thus, Equation (29) yields

$$\eta_\alpha^0 = \frac{\partial^\alpha \eta}{\partial t^\alpha} + (\eta_u - \alpha D_t(\tau))\frac{\partial^\alpha u}{\partial t^\alpha} - u\frac{\partial^\alpha \eta_u}{\partial t^\alpha} + \mu$$
$$+ \sum_{n=1}^\infty \left[\binom{\alpha}{n}\frac{\partial^\alpha \eta_u}{\partial t^\alpha} - \binom{\alpha}{n+1}D_t^{n+1}(\tau)\right] D_t^{\alpha-n}(u) - \sum_{n=1}^\infty \binom{\alpha}{n} D_t^n(\xi) D_t^{\alpha-n} u_x,$$

and

$$\vartheta_\alpha^0 = \frac{\partial^\alpha \vartheta}{\partial t^\alpha} + (\vartheta_v - \alpha D_t(\tau))\frac{\partial^\alpha v}{\partial t^\alpha} - u\frac{\partial^\alpha \vartheta_v}{\partial t^\alpha} + \lambda$$
$$+ \sum_{n=1}^{\infty}\left[\binom{\alpha}{n}\frac{\partial^\alpha \vartheta_v}{\partial t^\alpha} - \binom{\alpha}{n+1}D_t^{n+1}(\tau)\right]D_t^{\alpha-n}(v) - \sum_{n=1}^{\infty}\binom{\alpha}{n}D_t^n(\xi)D_t^{\alpha-n}v_x \quad (32)$$

4.2. Lie Symmetry

By third prolongation in Equations (1) and (2), we can obtain infinitesimals:

$$\begin{aligned}\xi &= \alpha x c_2 + c_1,\\ \tau &= 3tc_2,\\ \eta &= -2u\alpha c_2,\\ \vartheta &= -v\alpha c_2.\end{aligned} \quad (33)$$

Lie algebra corresponding to infinitesimal symmetry of governing system is spanned by

$$\mathbf{V}_1 = \frac{\partial}{\partial x} \quad (34)$$

$$\mathbf{V}_2 = x\alpha\frac{\partial}{\partial x} + 3t\frac{\partial}{\partial t} - 2u\alpha\frac{\partial}{\partial u} - v\alpha\frac{\partial}{\partial v} \quad (35)$$

Now, corresponding to Equations (1) and (2), we have following infinitesimal generators given as [7,8]

$$\mathbf{V} = c_1\mathbf{V}_1 + c_2\mathbf{V}_2$$

4.3. Similarity Reduction

Case 2: The following characteristic equation can be obtained by using the infinitesimal generator in Equation (35), given as

$$\frac{dx}{x\alpha} = \frac{dt}{3t} = -\frac{du}{2u\alpha} = -\frac{dv}{v\alpha} \quad (36)$$

After solving Equation (36), the following similarity variable can be obtained, given as

$$X = xt^{\frac{-\alpha}{3}} \quad (37)$$

$$u = F(X)t^{\frac{-2\alpha}{3}} \quad (38)$$

$$v = G(X)t^{\frac{-\alpha}{3}} \quad (39)$$

Theorem 1. *The transformation (38) and (39) reduces Equations (1) and (2) to the following form of Ordinary differential equations (ODEs) given as:*

$$\left(P_{\frac{3}{\alpha}}^{1-\frac{5\alpha}{3},\ \alpha}F\right)(X) + F_{XXX} + \frac{3}{2}GG_{XXX} + \frac{9}{2}G_XG_{XX} - 6\,FF_X - 6FGG_X - \frac{3}{2}F_XG^2 = 0 \quad (40)$$

$$\left(P_{\frac{3}{\alpha}}^{1-\frac{4\alpha}{3},\ \alpha}G\right)(X) + G_{XXX} - 6\,GF_X - 6FG_X - \frac{15}{2}G_XG^2 = 0 \quad (41)$$

with the Erdélyi-Kober operator $P_\beta^{\tau,\alpha}$:

$$\left(P_\beta^{\tau,\alpha}F\right) := \prod_{j=0}^{n-1}\left(\tau + j - \frac{1}{\beta}X\frac{d}{dX}\right)\left(K_\beta^{\tau+\alpha,n-\alpha}F\right)(X) \quad (42)$$

and
$$\left(P_\beta^{\tau,\alpha}G\right) := \prod_{j=0}^{n-1}\left(\tau + j - \frac{1}{\beta}X\frac{d}{dX}\right)\left(K_\beta^{\tau+\alpha,n-\alpha}G\right)(X) \qquad (43)$$

where, the Erdélyi-Kober fractional integral operator can be expressed as:

$$\left(K_\beta^{\tau+\alpha,n-\alpha}F\right)(X) := \begin{cases} \frac{1}{\Gamma(\alpha)}\int_1^\infty (u-1)^{\alpha-1}u^{-(\tau+\alpha)}F\left(Xu^{\frac{1}{\beta}}\right)du, & \alpha > 0, \\ F(X), & \alpha = 0. \end{cases} \qquad (44)$$

and

$$\left(K_\beta^{\tau+\alpha,n-\alpha}G\right)(X) := \begin{cases} \frac{1}{\Gamma(\alpha)}\int_1^\infty (u-1)^{\alpha-1}u^{-(\tau+\alpha)}G\left(Xu^{\frac{1}{\beta}}\right)du, & \alpha > 0, \\ G(X), & \alpha = 0. \end{cases} \qquad (45)$$

and

$$n = \begin{cases} [\alpha] + 1, & \alpha \in \mathcal{N}, \\ \alpha, & \alpha \notin \mathcal{N}. \end{cases} \qquad (46)$$

4.4. Conservation Laws of Time-Fractional Coupled JM Equations

Let us consider the following conservation vectors viz. C^1 and C^2 for the Equations (1) and (2), which satisfies the conservation equations expressed as:

$$[D_t(C^1) + D_x(C^2)]_{(1.1),\ (1.2)} = 0 \qquad (47)$$

A Lagrangian of Equations (1) and (2) is:

$$L = \omega(x,t)(D_t^\alpha u + u_{xxx} + \tfrac{3}{2}vv_{xxx} + \tfrac{9}{2}v_x v_{xx} - 6uu_x - 6uvv_x - \tfrac{3}{2}u_x v^2) \\ + \gamma(x,t)(D_t^\alpha v + v_{xxx} - 6u_x v - 6uv_x - \tfrac{15}{2}v_x v^2) \qquad (48)$$

where, γ and ω are dependent variables.

By considering Equation (48), the action integral can be defined as:

$$\int_0^t \int_\Omega L(x,\ t,\ u,\ v,\ \omega,\ \gamma,\ D_t^\alpha u,\ u_x,\ u_{xxx},\ D_t^\alpha v,\ v_x, v_{xxx})dx\ dt \qquad (49)$$

The Euler-Lagrangian operator is given by

$$\frac{\delta}{\delta u} = \frac{\partial}{\partial u} + (D_t^\alpha)^* \frac{\partial}{\partial D_t^\alpha u} - D_x \frac{\partial}{\partial u_x} - D_x^3 \frac{\partial}{\partial u_{xxx}} \qquad (50)$$

and

$$\frac{\delta}{\delta v} = \frac{\partial}{\partial v} + (D_t^\alpha)^* \frac{\partial}{\partial D_t^\alpha v} - D_x \frac{\partial}{\partial v_x} - D_x^2 \frac{\partial}{\partial v_{xx}} - D_x^3 \frac{\partial}{\partial v_{xxx}} \qquad (51)$$

where $(D_t^\alpha)^* = (-1)^n {}_t I_T^{n-\alpha} D_t^n$ is the adjoint operator of D_t^α.

Euler Lagrange equations:

$$\frac{\delta L}{\delta u} = 0,\ \text{and}\ \frac{\delta L}{\delta v} = 0 \qquad (52)$$

Considering the case of the independent variables t, x and the dependent variables $v(x,t)$, $u(x,t)$, we have

$$\overline{X} + D_t(\tau)I + D_x(\xi)I = W_1 \frac{\delta}{\delta u} + W_2 \frac{\delta}{\delta v} + D_t C^1 + D_x C^2 \qquad (53)$$

where $\frac{\delta}{\delta u}$, $\frac{\delta}{\delta v}$ are the Euler-Lagrange operators and I is the identity operator, C^1 and C^2 are the conserved vectors, and

So $\overline{X}$ is given as

$$\overline{X} = \xi \frac{\partial}{\partial x} + \tau \frac{\partial}{\partial t} + \eta \frac{\partial}{\partial u} + \vartheta \frac{\partial}{\partial v} + \eta_\alpha^0 \frac{\partial}{\partial _0 D_t^\alpha u} + \vartheta_\alpha^0 \frac{\partial}{\partial _0 D_t^\alpha v} \\ + \eta^x \frac{\partial}{\partial u_x} + \eta^{xxx} \frac{\partial}{\partial u_{xxx}} + \vartheta^x \frac{\partial}{\partial v_x} + \vartheta^{xx} \frac{\partial}{\partial v_{xx}} + \vartheta^{xxx} \frac{\partial}{\partial v_{xxx}}$$ (54)

Lie characteristic function W_1 and W_2 are:

$$W_1 = \eta - \tau u_t - \xi u_x$$
$$W_2 = \gamma - \tau v_t - \xi v_x$$

Here, for $\mathbf{V}_1$, we have following conserved vectors

$$W_1 = -u_x$$
$$W_2 = -v_x$$ (55)

Here, for $\mathbf{V}_2$, we have following conserved vectors

$$W_1 = -2u\alpha - x\alpha u_x - 3tu_t$$
$$W_2 = -v\alpha - x\alpha v_x - 3tv_t$$ (56)

In case of RL fractional differentiation in Equations (1) and (2), the components of the conserved vector can be written as follows:

For $W_1 = -2u\alpha - x\alpha u_x - 3tu_t$ and $W_2 = -v\alpha - x\alpha v_x - 3tv_t$, we have

$$C^1 = \tau L + {_0}D_t^{\alpha-1}(W_1) \frac{\partial L}{\partial _0 D_t^\alpha u} + J\left(W_1, D_t \frac{\partial L}{\partial _0 D_t^\alpha u}\right) + {_0}D_t^{\alpha-1}(W_2) \frac{\partial L}{\partial _0 D_t^\alpha v} + J\left(W_2, D_t \frac{\partial L}{\partial _0 D_t^\alpha v}\right),$$
$$= \omega \, {_0}D_t^{\alpha-1}(-2u\alpha - x\alpha u_x - 3tu_t) + J((-2u\alpha - x\alpha u_x - 3tu_t), \omega_t) \\ + \gamma_0 D_t^{\alpha-1}(-v\alpha - x\alpha v_x - 3tv_t) + J((-v\alpha - x\alpha v_x - 3tv_t), \gamma_t).$$ (57)

$$C^2 = \xi L + W_1\left[\frac{\partial L}{\partial u_x} + D_x D_x\left(\frac{\partial L}{\partial u_{xxx}}\right)\right] + W_2\left[\frac{\partial L}{\partial v_x} - D_x\left(\frac{\partial L}{\partial v_{xx}}\right) + D_x D_x\left(\frac{\partial L}{\partial v_{xxx}}\right)\right]$$
$$+ D_x(W_1)\left[-D_x\left(\frac{\partial L}{\partial u_{xxx}}\right)\right] + D_x(W_2)\left[\frac{\partial L}{\partial v_{xx}} - D_x\left(\frac{\partial L}{\partial v_{xxx}}\right)\right] + D_x D_x(W_1)\left(\frac{\partial L}{\partial u_{xxx}}\right) + D_x D_x(W_2)\left(\frac{\partial L}{\partial v_{xxx}}\right)$$
$$= \tfrac{1}{2}((4\alpha v_x \gamma_x + 6\alpha u_x \omega_x + 9tv_t v_x \omega_x + 3x\alpha v_x^2 \omega_x + 6t\omega_x u_{xt} + 6t\gamma_x v_{xt} + 9tv\omega_x v_{xt}$$
$$+ 2x\alpha(\omega_x u_{xx} + \gamma_x v_{xx}) + 3x\alpha v\omega_x v_{xx} - 2\alpha v \gamma_{xx} - 6tv_t \gamma_{xx} - 2x\alpha v_x \gamma_{xx} - 4\alpha u\omega_{xx}$$
$$- 3\alpha v^2 \omega_{xx} - 6tu_t \omega_{xx} - 9tv_x \omega_{xx} - 2x\alpha u_x \omega_{xx} + vv_x(9\alpha\omega_x - 3x\alpha\omega_{xx}))$$
$$+ \gamma(36\alpha uv + 15\alpha v^3 + 12v(3tu_t + x\alpha u_x) + 12u(3tv_t + x\alpha v_x) + 15v^2(3tv_t + x\alpha v_x)$$
$$- 6\alpha v_{xx} - 6tv_{xxt} - 2x\alpha v_{xxx}) + \omega(24\alpha u^2 + 18\alpha uv^2 + 12u(3tu_t + x\alpha u_x)$$
$$+ 3v^2(3tu_t + x\alpha u_x) - 12\alpha v_x^2 + 12uv(3tv_t + x\alpha v_x) - 18tv_x v_{xt} - 8\alpha u_{xx} - 12\alpha vv_{xx}$$
$$- 9tv_t v_{xx} - 9x\alpha v_x v_{xx} - 6tu_{xxt} - 9tvv_{xxt} - 2x\alpha u_{xxx} - 3x\alpha vv_{xxx}))$$ (58)

5. Conclusions

Fractional logistic function technique is proposed for soliton solutions of fractional JM system. Numerical simulation for solutions has been shown for analyzing the physical nature of obtained solutions. Moreover, Lie group analysis technique is proposed for investigation of symmetry properties and conservation laws for fractional Jaulent-Miodek system. Conservation laws for the system are acquired by new theorem and formal Lagrangian. These analyses are relatively new and reliable for finding exact solutions and constructing conservation laws with generating similarity solutions for the FPDEs. Furthermore, this method enriches the solution of the equations, which is of great significance for study of the FPDEs.

Author Contributions: Methodology, S.S.; validation, formal analysis, S.S.R.; software, investigation, M.A.M.A.; data curation, writing, original draft preparation, S.S.; writing, review and editing, M.I.; visualization, Y.-M.C. All authors have read and agreed to the published version of the manuscript.

Funding: The work was supported by the Natural Science Foundation of China (Grant Nos. 61673169, 11301127, 11701176, 11626101, 11601485).

Conflicts of Interest: The authors declare no conflict of interest.

References

1. Podlubny, I. *Fractional Differential Equation*; Academic Press: New York, NY, USA, 1999.
2. Miller, K.S.; Ross, B. *An Introduction to the Fractional Calculus and Fractional Differential Equations*; Wiley: New York, NY, USA, 1993; pp. 1–39. [CrossRef]
3. Samko, S.G.; Kilbas, A.A.; Marichev, O.I. *Fractional Integrals and Derivatives: Theory and Applications*; Taylor and Francis: London, UK, 2002.
4. Saha Ray, S.; Sahoo, S. *Generalized Fractional Order Differential Equations Arising in Physical Models*; Chapman and Hall/CRC: London, UK, 2018; p. 314.
5. Saha Ray, S. Dispersive Optical Solitons of Time-Fractional Schrödinger–Hirota Equation in Nonlinear Optical Fibers. *Phys. A Stat. Mech. Its Appl.* **2020**, *537*, 122619. [CrossRef]
6. Sadat, R.; Kassem, M.M. Lie Analysis and Novel Analytical Solutions for the Time-Fractional Coupled Whitham–Broer–Kaup Equations. *Int. J. Appl. Comput. Math* **2019**, *5*, 28. [CrossRef]
7. Olver, P.J. *Applications of Lie Groups to Differential Equations*; Springer Nature: New York, NY, USA, 1993. [CrossRef]
8. Oliveri, F. Lie Symmetries of Differential Equations: Classical Results and Recent Contributions. *Symmetry* **2010**, *2*, 658–706. [CrossRef]
9. Nadjafikhah, M.; Shirvani-Sh, V. Lie Symmetry Analysis of Kudryashov-Sinelshchikov Equation. *Math. Probl. Eng.* **2011**, *2011*, 1–9. [CrossRef]
10. Liu, H.; Li, J.; Zhang, Q. Lie Symmetry Analysis and Exact Explicit Solutions for General Burgers' Equation. *J. Comput. Appl. Math.* **2009**, *228*, 1–9. [CrossRef]
11. Sahoo, S.; Garai, G.; Saha Ray, S. Lie Symmetry Analysis for Similarity Reduction and Exact Solutions of Modified KdV–Zakharov–Kuznetsov Equation. *Nonlinear Dyn.* **2017**, *87*, 1995–2000. [CrossRef]
12. Dorodnitsyn, V.; Winternitz, P. Lie Point Symmetry Preserving Discretizations for Variable Coefficient Korteweg–De Vries Equations. *Nonlinear Dyn.* **2000**, *22*, 49–59. [CrossRef]
13. Baumann, G. *Symmetry Analysis of Differential Equations with Mathematica*; Springer: New York, NY, USA, 2000.
14. Sahoo, S.; Saha Ray, S. Lie Symmetry Analysis and Exact Solutions of (3 + 1) Dimensional Yu–Toda–Sasa–Fukuyama Equation in Mathematical Physics. *Comput. Math. Appl.* **2017**, *73*, 253–260. [CrossRef]
15. Atangana, A.; Baleanu, D. Nonlinear Fractional Jaulent-Miodek and Whitham-Broer-Kaup Equations within Sumudu Transform. *Abstr. Appl. Anal.* **2013**, *2013*, 1–8. [CrossRef]
16. Sahoo, S.; Saha Ray, S. New Solitary Wave Solutions of Time-Fractional Coupled Jaulent–Miodek Equation by Using Two Reliable Methods. *Nonlinear Dyn.* **2016**, *85*, 1167–1176. [CrossRef]
17. Şenol, M.; Iyiola, O.S.; Daei Kasmaei, H.; Akinyemi, L. Efficient Analytical Techniques for Solving Time-Fractional Nonlinear Coupled Jaulent–Miodek System with Energy-Dependent Schrödinger Potential. *Adv. Differ. Equ.* **2019**. [CrossRef]
18. Jaulent, M.; Miodek, I. Nonlinear Evolution Equations Associated with 'Energy–Dependent Schrödinger Potentials'. *Lett. Math. Phys.* **1976**, *1*, 243–250. [CrossRef]
19. Matsuno, Y. Reduction of Dispersionless Coupled Korteweg–De Vries Equations to the Euler–Darboux Equation. *J. Math. Phys.* **2001**, *42*, 1744–1760. [CrossRef]
20. Xu, G. N-fold Darboux Transformation of the Jaulent-Miodek Equation. *Appl. Math.* **2014**, *5*, 2657–2663. [CrossRef]
21. Ruan, H.; Lou, S. New Symmetries of the Jaulent-Miodek Hierarchy. *J. Phys. Soc. Jpn.* **1993**, *62*, 1917–1921. [CrossRef]
22. Wazwaz, A.M. The Tanh–Coth and the Sech Methods for Exact Solutions of the Jaulent–Miodek Equation. *Phys. Lett. A* **2007**, *366*, 85–90. [CrossRef]
23. Rashidi, M.M.; Domairry, G.; Dinarvand, S. The Homotopy Analysis Method for Explicit Analytical Solutions of Jaulent-Miodek Equations. *Numer. Methods Part. Differ. Equ.* **2009**, *25*, 430–439. [CrossRef]

24. He, J.H.; Zhang, L.N. Generalized Solitary Solution and Compacton-Like Solution of the Jaulent–Miodek Equations Using the Exp-Function Method. *Phys. Lett. A* **2008**, *372*, 1044–1047. [CrossRef]
25. Zayed, E.M.E.; Rahman, H.M.A. The Extended Tanh-Method for Finding Traveling Wave Solutions of Nonlinear Evolution Equations. *Appl. Math.* **2010**, *10*, 235–245.
26. Malfliet, W. Solitary Wave Solutions of Nonlinear Wave Equations. *Am. J. Phys.* **1992**, *60*, 650–654. [CrossRef]
27. Atangana, A.; Cloot, A.H. Stability and Convergence of the Space Fractional Variable-Order Schrödinger Equation. *Adv. Differ. Equ.* **2013**, *2013*. [CrossRef]
28. Das, G.C.; Sarma, J.; Uberoi, C. Explosion of Soliton in a Multicomponent Plasma. *Phys. Plasmas* **1997**, *4*, 2095–2100. [CrossRef]
29. Hong, T.; Wang, Y.-z.; Huo, Y.-s. Bogoliubov Quasiparticles Carried by Dark Solitonic Excitations in Nonuniform Bose-Einstein Condensates. *Chin. Phys. Lett.* **1998**, *15*, 550–552. [CrossRef]
30. Lou, S.-y. A Direct Perturbation Method: Nonlinear Schrödinger Equation with Loss. *Chin. Phys. Lett.* **1999**, *16*, 659–661. [CrossRef]
31. Ma, W.-X.; Li, C.-X.; He, J. A Second Wronskian Formulation of the Boussinesq Equation. *Nonlinear Anal. Theory Methods Appl.* **2009**, *70*, 4245–4258. [CrossRef]
32. Zhang, J.-F. Multiple Soliton Solutions of the Dispersive Long-Wave Equations. *Chin. Phys. Lett.* **1999**, *16*, 4–5. [CrossRef]
33. Wang, H.; Xia, T.-C. The Fractional Supertrace Identity and Its Application to the Super Jaulent–Miodek Hierarchy. *Commun. Nonlinear Sci. Numer. Simul.* **2013**, *18*, 2859–2867. [CrossRef]
34. Abedl-Rady, A.S.; Rida, S.Z.; Arafa, A.A.M.; Abedl-Rahim, H.R. Fractional Physical Models via Natural Transform. *IOSR J. Math.* **2016**, *12*, 31–37.
35. Gupta, A.K.; Saha Ray, S. An Investigation with Hermite Wavelets for Accurate Solution of Fractional Jaulent–Miodek Equation Associated with Energy-Dependent Schrödinger Potential. *Appl. Math. Comput.* **2015**, *270*, 458–471. [CrossRef]
36. Sahoo, S.; Ray, S.S. A New Method for Exact Solutions of Variant Types of Time-Fractional Korteweg-De Vries Equations in Shallow Water Waves. *Math. Methods Appl. Sci.* **2017**, *40*, 106–114. [CrossRef]
37. Ibragimov, N.H. A New Conservation Theorem. *J. Math. Anal. Appl.* **2007**, *333*, 311–328. [CrossRef]
38. Yaşar, E. On the Conservation Laws and Invariant Solutions of the mKdV Equation. *J. Math. Anal. Appl.* **2010**, *363*, 174–181. [CrossRef]
39. Hu, J.; Ye, Y.; Shen, S.; Zhang, J. Lie Symmetry Analysis of the Time Fractional KdV-Type Equation. *Appl. Math. Comput.* **2014**, *233*, 439–444. [CrossRef]
40. Lukashchuk, S.Y. Conservation Laws for Time-Fractional Subdiffusion and Diffusion-Wave Equations. *Nonlinear Dyn.* **2015**, *80*, 791–802. [CrossRef]
41. Gazizov, R.K.; Ibragimov, N.H.; Lukashchuk, S.Y. Nonlinear Self-Adjointness, Conservation Laws and Exact Solutions of Time-Fractional Kompaneets Equations. *Commun. Nonlinear Sci. Numer. Simul.* **2015**, *23*, 153–163. [CrossRef]
42. Hu, M.S.; Baleanu, D.; Yang, X.J. One-Phase Problems for Discontinuous Heat Transfer in Fractal Media. *Math. Probl. Eng.* **2013**, *2013*, 358473. [CrossRef]
43. Yang, X.J. The Zero-Mass Renormalization Group Differential Equations and Limit Cycles in Non-Smooth Initial Value Problems. *Prespacetime J.* **2012**, *3*, 913–923.
44. Su, W.H.; Yang, X.J.; Jafari, H.; Baleanu, D. Fractional Complex Transform Method for Wave Equations on Cantor sets Within Local Fractional Differential Operator. *Adv. Differ. Equ.* **2013**, *2013*, 1–8. [CrossRef]
45. Bekir, A.; Güner, Ö.; Cevikel, A.C. Fractional Complex Transform and Exp-Function Methods for Fractional Differential Equations. *Abstr. Appl. Anal.* **2013**, *2013*, 426462. [CrossRef]
46. Kudryashov, N.A. Polynomials in Logistic Function and Solitary Waves of Nonlinear Differential Equations. *Appl. Math. Comput.* **2013**, *219*, 9245–9253. [CrossRef]
47. Kudryashov, N.A. Logistic Function as Solution of Many Nonlinear Differential Equations. *Appl. Math. Model.* **2015**, *39*, 5733–5742. [CrossRef]
48. Senthilvelan, M. On the Extended Applications of Homogenous Balance Method. *Appl. Math. Comput.* **2001**, *123*, 381–388. [CrossRef]
49. Sirisubtawee, S.; Koonprasert, S. Exact Traveling Wave Solutions of Certain Nonlinear Partial Differential Equations Using the G'/G2-Expansion Method. *Adv. Math. Phys.* **2018**, *2018*, 1–15. [CrossRef]

50. Djordjevic, V.D.; Atanackovic, T.M. Similarity Solutions to Nonlinear Heat Conduction and Burgers/Korteweg–deVries Fractional Equations. *J. Comput. Appl. Math.* **2008**, *222*, 701–714. [CrossRef]
51. Gazizov, R.K.; Kasatkin, A.A.; Lukashchuk, S.Y. Symmetry Properties of Fractional Diffusion Equations. *Phys. Scr.* **2009**, *T136*, 014016. [CrossRef]
52. Sahadevan, R.; Bakkyaraj, T. Invariant Analysis of Time Fractional Generalized BURGERS and Korteweg–De Vries Equations. *J. Math. Anal. Appl.* **2012**, *393*, 341–347. [CrossRef]

© 2020 by the authors. Licensee MDPI, Basel, Switzerland. This article is an open access article distributed under the terms and conditions of the Creative Commons Attribution (CC BY) license (http://creativecommons.org/licenses/by/4.0/).

Article

On Nonlinear Fractional Difference Equation with Delay and Impulses

Rujira Ouncharoen [1], Saowaluck Chasreechai [2,*] and Thanin Sitthiwirattham [3,*]

[1] Research Center in Mathematics and Applied Mathematics, Department of Mathematics, Faculty of Science, Chiang Mai University, Chiang Mai 50200, Thailand; rujira.o@cmu.ac.th
[2] Department of Mathematics, Faculty of Applied Science, King Mongkut's University of Technology North Bangkok, Bangkok 10800, Thailand
[3] Mathematics Department, Faculty of Science and Technology, Suan Dusit University, Bangkok 10300, Thailand
* Correspondence: saowaluck.c@sci.kmutnb.ac.th (S.C.); thanin_sit@dusit.ac.th (T.S.)

Received: 12 May 2020; Accepted: 5 June 2020; Published: 8 June 2020

Abstract: In this paper, we establish the existence results for a nonlinear fractional difference equation with delay and impulses. The Banach and Schauder's fixed point theorems are employed as tools to study the existence of its solutions. We obtain the theorems showing the conditions for existence results. Finally, we provide an example to show the applicability of our results.

Keywords: fractional difference equations; delay; impulses; existence

JEL Classification: 39A05; 39A12

1. Introduction

Discrete fractional calculus studies have been an interesting field of present day, because some real-world phenomena are described by using fractional difference operators (see papers [1–3] and the references therein). Basic knowledge of fractional difference calculus can be found in [4]. The extension of this field can be found in [5–37] and references cited therein.

For the development of the fractional difference equations theory, which is the discrete case of fractional differential equations, there are still few publications. However, there are some recent papers studying fractional difference equations with delay. In 2017, Kaewwisetkul et al. [38] studied boundary value problems for Caputo fractional functional difference equations with delay. In 2018, Wu et al. [39] proposed the finite-time stability of discrete fractional delay systems, Alzabut et al. [40] studied nonlinear delay fractional difference equations with applications on the discrete fractional Lotka–Volterra competition model, Alzabut et al. [41] investigated the application on the uniqueness of solutions for nonlinear delay fractional difference system, and Luo et al. [42] considered the uniqueness and finite-time stability of solutions for a class of nonlinear fractional delay difference systems.

In particular, the fractional difference equations with delay and impulses have not been studied extensively. In 2018, Wu et al. [43] studied a linear fractional delay difference equations with impulse. These results are incentives for research. In this paper, we propose a nonlinear fractional difference equation with delay and impulses of the form:

$$\begin{aligned}
&\Delta_C^\alpha u(t) = F\left[t+\alpha-1, u_{t+\alpha-1}, \Delta^\beta u(t+\alpha-\beta)\right], \quad t \in \mathbb{N}_{0,T},\ t+\alpha-1 \neq t_k \\
&u(t_k) = I_k\left(u_{t_k-1}\right), \quad k=1,2,\dots,p,\ t_{k+1}-t_k \geq 2, \\
&u(t+\alpha-1) = \psi(t+\alpha-1), \quad t \in \mathbb{N}_{-r,0},\ r \in \mathbb{N}_{0,T+1},
\end{aligned} \qquad (1)$$

where $\mathbb{N}_{0,T} := \{0, 1, \ldots, T\}$, $\alpha, \beta \in (0,1)$, $\Delta u(t_k) = u(t_k + 1) - u(t_k)$, $t_0 = \alpha - 1 < t_1 < t_2 < \ldots < t_p < T + \alpha$, $F \in C(\mathbb{N}_{\alpha-1, T+\alpha} \times C_r \times \mathbb{R}, \mathbb{R})$, $I_k : C_r \to \mathbb{R}$ and ψ is an element of the space:

$$C_r^+(\alpha - 1) := \left\{ \psi \in C_r : \psi(\alpha - 1) = 0,\ \Delta_C^\beta \psi(s - \beta + 1) = 0,\ s \in \mathbb{N}_{\alpha - r - 1, \alpha - 1} \right\}.$$

For $r \in \mathbb{N}_{0, T+1}$, let C_r be the Banach space of all continuous functions $\psi : \mathbb{N}_{\alpha-r-1, \alpha-1} \to \mathbb{R}$ with the norm:

$$\|\psi\|_{C_r} = \max_{s \in \mathbb{N}_{\alpha-r-1, \alpha-1}} |\psi(s)|.$$

If $u : \mathbb{N}_{\alpha-r-1, \alpha-1} \to \mathbb{R}$, then for any $t \in \mathbb{N}_{\alpha-1, T+\alpha}$, we define the element u_t of C_r as,

$$u_t(\theta) = u(t + \theta) \text{ for } \theta \in \mathbb{N}_{-r, 0}.$$

We aim to prove the existence results to the problem of Equation (1) by using the Banach and Schauder's fixed point theorems. Finally, we present an example in the last section.

2. Preliminaries

In this section, we recall some notations, definitions, and lemmas used in the main results.

Definition 1. *The generalized falling function is defined by:*

$$t^{\underline{\alpha}} := \frac{\Gamma(t+1)}{\Gamma(t+1-\alpha)}.$$

If $t + 1 - \alpha$ is a pole of the Gamma function and $t + 1$ is not a pole, then $t^{\underline{\alpha}} = 0$.

Definition 2. *For $\alpha > 0$ and f defined on $\mathbb{N}_a := \{a, a+1, \ldots\}$, the α-order fractional sum of f is defined by:*

$$\Delta^{-\alpha} f(t) := \frac{1}{\Gamma(\alpha)} \sum_{s=a}^{t-\alpha} (t - \sigma(s))^{\underline{\alpha-1}} f(s),$$

where $t \in \mathbb{N}_{a+\alpha}$ and $\sigma(s) = s + 1$.

Definition 3. *For $\alpha > 0$, $N \in \mathbb{N}$ is satisfied with $0 \leq N - 1 < \alpha < N$ and f defined on $\mathbb{N}_a$, the α-order Riemann–Liouville fractional difference of f is defined by:*

$$\Delta^\alpha f(t) := \Delta^N \Delta^{-(N-\alpha)} f(t) = \frac{1}{\Gamma(-\alpha)} \sum_{s=a}^{t+\alpha} (t - \sigma(s))^{\underline{-\alpha-1}} f(s),$$

where $t \in \mathbb{N}_{a+N-\alpha}$. The α-order Caputo fractional difference of f is defined by:

$$\Delta_C^\alpha f(t) := \Delta^{-(N-\alpha)} \Delta^N f(t) = \frac{1}{\Gamma(N-\alpha)} \sum_{s=a}^{t-(N-\alpha)} (t - \sigma(s))^{\underline{N-\alpha-1}} \Delta^N f(s),$$

where $t \in \mathbb{N}_{a+N-\alpha}$. If $\alpha = N$, then $\Delta^\alpha f(t) = \Delta_C^\alpha f(t) = \Delta^N f(t)$.

Lemma 1. *[5] Assume that $\alpha > 0$ and f defined on $\mathbb{N}_a$. Then,*

$$\Delta^{-\alpha} \Delta_C^\alpha y(t) = y(t) + C_0 + C_1 (t-a)^{\underline{1}} + C_2 (t-a)^{\underline{2}} + \ldots + C_{N-1} (t-a)^{\underline{N-1}},$$

for some $C_i \in \mathbb{R}$, $0 \leq i \leq N-1$ and $0 \leq N - 1 < \alpha \leq N$.

Next, we aim to find a solution of the linear variant of the mixed problem in Equation (1) as follows.

Lemma 2. *Let $\alpha \in (0,1)$, $h \in C(\mathbb{N}_{\alpha-1,T+\alpha}, \mathbb{R})$, $I_k : C_r \to \mathbb{R}$ and $\psi \in C_r^+(\alpha - 1)$ be given. Then the problem*

$$\begin{aligned} \Delta_C^\alpha u(t) &= h(t + \alpha - 1), \quad t \in \mathbb{N}_{0,T} := \{0, 1, \ldots, T\}, \ t + \alpha - 1 \neq t_k \\ u(t_k) &= I_k\left(u_{t_k-1}\right), \quad k = 1, 2, \ldots, p, \ t_{k+1} - t_k \geq 2, \\ u(t + \alpha - 1) &= \psi(t + \alpha - 1), \quad t \in \mathbb{N}_{-r,0}, \ r \in \mathbb{N}_{0,T+1}, \end{aligned} \quad (2)$$

has the unique solution which is in a form:

$$u(t) = \begin{cases} \dfrac{1}{\Gamma(\alpha)} \sum_{s=t_0}^{t-1} (t-s+\alpha-2)^{\underline{\alpha-1}} h(s), & t \in \mathbb{N}_{t_0,t_1} \\ \sum_{i=1}^{k} I_i\left(u_{t_i-1}\right) + \dfrac{1}{\Gamma(\alpha)} \sum_{i=1}^{k} \sum_{s=t_{i-1}}^{t_i-1} (t_i-s+\alpha-2)^{\underline{\alpha-1}} h(s) \\ \quad + \dfrac{1}{\Gamma(\alpha)} \sum_{s=t_k}^{t-1} (t-s+\alpha-2)^{\underline{\alpha-1}} h(s), & t \in \mathbb{N}_{t_k+1,t_{k+1}} \\ \varphi(t), & t \in \mathbb{N}_{\alpha-r-1,\alpha-1} \end{cases} \quad (3)$$

where $\Delta u(t_k) = u(t_k+1) - u(t_k)$, $t_0 = \alpha - 1 < t_1 < t_2 < \ldots < t_p < T + \alpha$.

Proof. For $t \in \mathbb{N}_{t_0,t_1}$, taking the fractional sum of order α for Equation (2) and from Lemma 1, we have:

$$u(t) = \varphi(\alpha - 1) + \dfrac{1}{\Gamma(\alpha)} \sum_{s=0}^{t-\alpha} (t - \sigma(s))^{\underline{\alpha-1}} h(s + \alpha - 1). \quad (4)$$

From $\varphi(\alpha - 1) = 0$, we can write Equation (4) as:

$$u(t) = \dfrac{1}{\Gamma(\alpha)} \sum_{s=t_0}^{t-1} (t - s + \alpha - 2)^{\underline{\alpha-1}} h(s). \quad (5)$$

By substituting $t = t_1$ into Equation (5), we have:

$$u(t_1) = \dfrac{1}{\Gamma(\alpha)} \sum_{s=t_0}^{t_1-1} (t_1 - s + \alpha - 2)^{\underline{\alpha-1}} h(s). \quad (6)$$

If $t \in \mathbb{N}_{t_1+1,t_2}$, then we get"

$$\begin{aligned} u(t) &= u(t_1 + 1) + \dfrac{1}{\Gamma(\alpha)} \sum_{s=t_1}^{t-1} (t - s + \alpha - 2)^{\underline{\alpha-1}} h(s) \\ &= \Delta u(t_1) + u(t_1) + \dfrac{1}{\Gamma(\alpha)} \sum_{s=t_1}^{t-1} (t - s + \alpha - 2)^{\underline{\alpha-1}} h(s). \end{aligned}$$

Substiting $u(t_1)$ from Equation (6) to above equation, we obtain:

$$u(t) = I_1(u_{t_1-1}) + \frac{1}{\Gamma(\alpha)} \sum_{s=t_0}^{t_1-1} (t_1 - s + \alpha - 2)^{\underline{\alpha-1}} h(s)$$
$$+ \frac{1}{\Gamma(\alpha)} \sum_{s=t_1}^{t-1} (t - s + \alpha - 2)^{\underline{\alpha-1}} h(s). \quad (7)$$

If $t \in \mathbb{N}_{t_2+1,t_3}$, then we have:

$$u(t) = u(t_2 + 1) + \frac{1}{\Gamma(\alpha)} \sum_{s=t_2}^{t-1} (t - s + \alpha - 2)^{\underline{\alpha-1}} h(s)$$
$$= \Delta u(t_2) + u(t_2) + \frac{1}{\Gamma(\alpha)} \sum_{s=t_2}^{t-1} (t - s + \alpha - 2)^{\underline{\alpha-1}} h(s).$$

Substituting $u(t_2)$ from Equation (7) to above equation, we obtain:

$$u(t) = I_2(u_{t_2-1}) + \left[I_1(u_{t_1-1}) + \frac{1}{\Gamma(\alpha)} \sum_{s=t_0}^{t_1-1} (t_1 - s + \alpha - 2)^{\underline{\alpha-1}} h(s) \right.$$
$$\left. + \frac{1}{\Gamma(\alpha)} \sum_{s=t_1}^{t_2-1} (t_2 - s + \alpha - 2)^{\underline{\alpha-1}} h(s) \right] + \frac{1}{\Gamma(\alpha)} \sum_{s=t_2}^{t-1} (t - s + \alpha - 2)^{\underline{\alpha-1}} h(s). \quad (8)$$

By using the recursive process, we obtain the solution $u(t)$ for $t \in \mathbb{N}_{t_k+1,t_{k+1}}$ ($k = 1, 2, ..., p$) as given by:

$$u(t) = \sum_{i=1}^{k} I_i(u_{t_i-1}) + \frac{1}{\Gamma(\alpha)} \sum_{i=1}^{k} \sum_{s=t_{i-1}}^{t_i-1} (t_i - s + \alpha - 2)^{\underline{\alpha-1}} h(s)$$
$$+ \frac{1}{\Gamma(\alpha)} \sum_{s=t_k}^{t-1} (t - s + \alpha - 2)^{\underline{\alpha-1}} h(s). \quad (9)$$

Obviously, for each $t \in \mathbb{N}_{\alpha-r-1,\alpha-1}$, we have $u(t) = \varphi(t)$. □

3. Existence and Uniqueness Result

In this section, we employ the Banach fixed point theorem to consider the existence and uniqueness result for the problem in Equation (1). Define the Banach space:

$$\mathcal{X} := \left\{ u : u \in C(\mathbb{N}_{\alpha-r-1,T+\alpha}, \mathbb{R}), \, \Delta^\beta u \in C(\mathbb{N}_{\alpha-\beta-r,T+\alpha-\beta+1}, \mathbb{R}), \, 0 < \beta < 1 \right\}$$

with the norm defined by:

$$\|u\|_{\mathcal{X}} = \|u\| + \|\Delta^\beta u\|, \quad (10)$$

where $\|u\| = \max_{t \in \mathbb{N}_{\alpha-r-1,T+\alpha}} |u(t)|$ and $\|\Delta^\beta u\| = \max_{t \in \mathbb{N}_{\alpha-r-1,T+\alpha}} |\Delta^\nu u(t - \beta + 1)|$.

In view of the definitions of u_t and ψ, we have:

$$u_{\alpha-1} = u_{\alpha-1}(\theta) = u(\theta + \alpha - 1) = \psi(\theta + \alpha - 1) \text{ for } \theta \in \mathbb{N}_{-r,0}. \quad (11)$$

Thus, we obtainL

$$u(t) = \psi(t) \text{ for } t \in \mathbb{N}_{\alpha-r-1, \alpha-1}. \tag{12}$$

Next, define an operator $\mathcal{T} : \mathcal{X} \to \mathcal{X}$ as:

$$(\mathcal{T}u)(t) := \begin{cases} \dfrac{1}{\Gamma(\alpha)} \sum_{s=t_0}^{t-1} (t-s+\alpha-2)^{\underline{\alpha-1}} F\left[s, u_s, \Delta^\beta u(s-\beta+1)\right], & t \in \mathbb{N}_{t_0, t_1} \\ \sum_{i=1}^{k} I_i\left(u_{t_i-1}\right) + \dfrac{1}{\Gamma(\alpha)} \sum_{i=1}^{k} \sum_{s=t_{i-1}}^{t_i-1} (t_i-s+\alpha-2)^{\underline{\alpha-1}} F\left[s, u_s, \Delta^\beta u(s-\beta+1)\right] \\ \quad + \dfrac{1}{\Gamma(\alpha)} \sum_{s=t_k}^{t-1} (t-s+\alpha-2)^{\underline{\alpha-1}} F\left[s, u_s, \Delta^\beta u(s-\beta+1)\right], & t \in \mathbb{N}_{t_k+1, t_{k+1}} \\ \varphi(t), & t \in \mathbb{N}_{\alpha-r-1, \alpha-1} \end{cases} \tag{13}$$

where $k = 1, 2, ..., p$, $t_{k+1} - t_k \geq 2$, $t_0 = \alpha - 1 < t_1 < t_2 < ... < t_p < T + \alpha$.

Firstly, we provide some basic knowledge that is used in this section as follows.

Definition 4. *A mapping S from a subset M of a Banach space X into X is called a contraction mapping (or simply a contraction) if there exists a positive number $\alpha < 1$ such that:*

$$\|S(x) - S(y)\|_X \leq \alpha \|x - y\|_X \text{ for all } x, y \in M.$$

Lemma 3. *[44] (Banach fixed point theorem) Let M be a closed subset of a Banach space X and let S be a contraction mapping from M into M. Then there exists a unique $z \in M$ such that $S(z) = z$.*

If one can prove that $\mathcal{T}$ has fixed point, we can conclude that the problem of Equation (1) has a solution.

Theorem 1. *Assume the following properties:*

(H1) *There exists a constant $\ell > 0$ such that:*

$$\left| F[t, u_1, u_2] - F[t, v_1, v_2] \right| \leq \ell \left(|u_1 - v_1| + |u_2 - v_2| \right),$$

for each $u_1, v_1 \in C_r$ and $u_2, v_2 \in \mathbb{R}$.

(H2) *There exists a constant $\lambda > 0$ such that*

$$\left| I_k\left(u_{t_{k-1}}\right) - I_k\left(v_{t_{k-1}}\right) \right| \leq \lambda |u_t - v_t|,$$

for each $u_t, v_t \in C_r$ and $k = 1, 2, ..., p$.

(H3) $\left[\dfrac{p(p+1)}{2} \lambda + \ell \left(\dfrac{p(p+1)}{2} + 1 \right) \dfrac{(T+\alpha)^{\underline{\alpha}}}{\Gamma(\alpha+1)} \right] \left[1 + \dfrac{(T+\alpha-\beta-1)^{\underline{-\beta}}}{\Gamma(1-\beta)} \right] < 1.$

Then, the problem of Equation (1) has an unique solution.

Proof. We will show that $\mathcal{T}$ is a contraction. Letting,

$$|\mathcal{H}(u-v)|(t) = \left| F\left[t, u_t, \Delta^\beta u(t-\beta+1)\right] - F\left[t, v_t, \Delta^\beta v(t-\beta+1)\right] \right|,$$

for each $t \in \mathbb{N}_{\alpha-1,T+\alpha}$, we obtain:

$$|(\mathcal{T}u)(t) - (\mathcal{T}v)(t)| \leq \sum_{i=1}^{p} |I_i(u_{t_i-1}) - I_i(v_{t_i-1})| + \frac{1}{\Gamma(\alpha)} \sum_{i=1}^{k} \sum_{s=t_{i-1}}^{t_i-1} (t_i - s + \alpha - 2)^{\underline{\alpha-1}} |\mathcal{H}(u-v)|(s)$$

$$+ \frac{1}{\Gamma(\alpha)} \sum_{s=t_k}^{t-1} (t - s + \alpha - 2)^{\underline{\alpha-1}} |\mathcal{H}(u-v)|(s)$$

$$\leq \frac{p(p+1)}{2} \lambda |u_t - v_t| + \frac{\ell \left(|u_t - v_t| + |\Delta^\beta u(t-\beta+1) - \Delta^\beta v(t-\beta+1)| \right)}{\Gamma(\alpha+1)} \times$$

$$\left\{ \left(\frac{p(p+1)}{2} + 1 \right) (T+\alpha)^{\underline{\alpha}} \right\}$$

$$\leq \left[\frac{p(p+1)}{2} \lambda + \ell \left(\frac{p(p+1)}{2} + 1 \right) \frac{(T+\alpha)^{\underline{\alpha}}}{\Gamma(\alpha+1)} \right] \|u - v\|_{\mathcal{X}}. \quad (14)$$

Taking the fractional difference of order β for Equation (13) and substituting $t = t - \beta + 1$, we get:

$$\left(\Delta^\beta \mathcal{T} u \right)(t - \beta + 1)$$

$$:= \begin{cases} \frac{1}{\Gamma(\alpha)\Gamma(-\beta)} \sum_{\zeta=t_0+1}^{t+1} \sum_{s=t_0}^{\zeta-1} (t-\beta-\zeta)^{\underline{-\beta-1}} (\zeta - s + \alpha - 2)^{\underline{\alpha-1}} \times \\ F[s, u_s, \Delta^\beta u(s-\beta+1)], & t \in \mathbb{N}_{t_0, t_1} \\ \frac{1}{\Gamma(-\beta)} \sum_{s=t_0}^{t+1} \sum_{i=1}^{k} (t-\beta-s)^{\underline{-\beta-1}} I_i(u_{t_i-1}) \\ + \frac{1}{\Gamma(\alpha)\Gamma(-\beta)} \sum_{\zeta=t_0}^{t+1} \sum_{i=1}^{k} \sum_{s=t_{i-1}}^{t_i-1} (t-\beta-\zeta)^{\underline{-\beta-1}} (t_i - s + \alpha - 2)^{\underline{\alpha-1}} \times \\ F[s, u_s, \Delta^\beta u(s-\beta+1)] \\ + \frac{1}{\Gamma(\alpha)\Gamma(-\beta)} \sum_{\zeta=t_0+1}^{t+1} \sum_{s=t_k}^{\zeta-1} (t-\beta-\zeta)^{\underline{-\beta-1}} (\zeta - s + \alpha - 2)^{\underline{\alpha-1}} \times \\ F[s, u_s, \Delta^\beta u(s-\beta+1)], & t \in \mathbb{N}_{t_k+1, t_{k+1}} \\ \frac{1}{\Gamma(-\beta)} \sum_{s=t_0-r}^{t+1} (t-\beta-s)^{\underline{-\beta-1}} \varphi(s), & t \in \mathbb{N}_{\alpha-r-1, \alpha-1} \end{cases} \quad (15)$$

For each $t \in \mathbb{N}_{\alpha-1,T+\alpha}$, we obtain:

$$\left| (\Delta^\beta \mathcal{T} u)(t-\beta+1) - (\Delta^\beta \mathcal{T} v)(t-\beta+1) \right|$$

$$\leq \frac{1}{\Gamma(-\beta)} \sum_{s=t_0}^{T+\alpha+1} (T+\alpha-\beta-s)^{\underline{-\beta-1}} \sum_{i=1}^{p} |I_i(u_{t_i-1}) - I_i(v_{t_i-1})| + \frac{1}{\Gamma(\alpha)\Gamma(-\beta)} \times$$

$$\sum_{\zeta=t_0}^{T+\alpha+1} \sum_{i=1}^{k} \sum_{s=t_{i-1}}^{t_i-1} (T+\alpha-\beta-\zeta)^{\underline{-\beta-1}} (t_i - s + \alpha - 2)^{\underline{\alpha-1}} |\mathcal{H}(u-v)|(s)$$

$$+ \frac{1}{\Gamma(\alpha)\Gamma(-\beta)} \sum_{\zeta=t_0+1}^{T+\alpha+1} \sum_{s=t_k}^{\zeta-1} (T+\alpha-\beta-\zeta)^{\underline{-\beta-1}} (\zeta - s + \alpha - 2)^{\underline{\alpha-1}} |\mathcal{H}(u-v)|(s)$$

$$\leq \frac{p(p+1)}{2} \lambda |u_t - v_t| + \frac{\ell \left(|u_t - v_t| + |\Delta^\beta u(t-\beta+1) - \Delta^\beta v(t-\beta+1)| \right)}{\Gamma(\alpha+1)} \times$$

$$\left\{ \left(\frac{p(p+1)}{2} + 1 \right) (T+\alpha)^{\underline{\alpha}} \right\}$$

$$\leq \frac{(T+\alpha-\beta-1)^{\underline{-\beta}}}{\Gamma(1-\beta)} \left[\frac{p(p+1)}{2} \lambda + \ell \left(\frac{p(p+1)}{2} + 1 \right) \frac{(T+\alpha)^{\underline{\alpha}}}{\Gamma(\alpha+1)} \right] \|u - v\|_{\mathcal{X}}. \quad (16)$$

Obviously, for each $t \in \mathbb{N}_{\alpha-r-1,\alpha-1}$, we get $\left|(\mathcal{T}u)(t) - (\mathcal{T}v)(t)\right| = 0$.
Therefore, we have:

$$\left\|(\mathcal{T}u) - (\mathcal{T}v)\right\|_{\mathcal{X}} \tag{17}$$
$$\leq \left[\frac{p(p+1)}{2}\lambda + \ell\left(\frac{p(p+1)}{2}+1\right)\frac{(T+\alpha)^{\alpha}}{\Gamma(\alpha+1)}\right]\left[1 + \frac{(T+\alpha-\beta-1)^{-\beta}}{\Gamma(1-\beta)}\right]\|x-y\|_{\mathcal{X}}.$$

By (H3), it implies that $\mathcal{T}$ is a contraction. Therefore, by Banach fixed point theorem, $\mathcal{T}$ has a fixed point which is a unique solution of the problem in Equation (1). □

4. Existence of at Least One Solution

In this section, we also present the existence of at least one solution of Equation (1) by using the Schauder's fixed point theorem. Firstly, we provide some basic knowledge that is used in this section as follows.

Lemma 4. [44] (Arzelá–Ascoli theorem) *A set of function in $C[a,b]$ with the sup norm, is relatively compact if and only it is uniformly bounded and equicontinuous on $[a,b]$.*

Lemma 5. [44] *A bounded set in $\mathbb{R}^n$ is relatively compact, a closed bounded set in $\mathbb{R}^n$ is compact.*

Lemma 6. [45] (Schauder's fixed point theorem) *Let (D,d) be a complete metric space, U be a closed convex subset of D, and $T : D \to D$ be the map such that the set $Tu : u \in U$ is relatively compact in D. Then the operator T has at least one fixed point $u^* \in U$: $Tu^* = u^*$.*

The following notations are defined for using in the sequel.

$$\Theta = \max_{t \in \mathbb{N}_{\alpha-r-1,\alpha-1}} \left\{\left(\frac{1}{2}p(p+1)+1\right)\frac{(T+\alpha)^{\alpha}}{\Gamma(\alpha+1)}\phi(t)\right\} \tag{18}$$

$$\tilde{\Theta} = \max_{t \in \mathbb{N}_{\alpha-r-1,\alpha-1}} \left\{\left(\frac{1}{2}p(p+1)+1\right)\frac{(T+\alpha)^{\alpha}(T+\alpha-\beta+1)^{-\beta}}{\Gamma(\alpha+1)\Gamma(1-\beta)}\phi(t)\right\} \tag{19}$$

$$Y = \left(\frac{1}{2}p(p+1)+1\right)\frac{(T+\alpha)^{\alpha}}{\Gamma(\alpha+1)}\left[1 + \frac{(T+\alpha-\beta+1)^{-\beta}}{\Gamma(1-\beta)}\right]. \tag{20}$$

Theorem 2. *Assume the following properties:*

(H4) *There exists a nonnegative function $\phi \in C\left(\mathbb{N}_{\alpha-1,T+\alpha}\right)$ such that:*

$$\left|F[t,x,y]\right| \leq \phi(t) + \lambda_1|x|^{\chi_1} + \lambda_2|y|^{\chi_2},$$

for each $x \in C_r$, $y \in \mathbb{R}$ where λ_1, λ_2 are negative constants and $0 < \chi_1, \chi_2 < 1$; or

(H5) *there exists a nonnegative function $\phi \in C\left(\mathbb{N}_{\alpha-3,T+\alpha}\right)$ such that:*

$$\left|F[t,x,y]\right| \leq \phi(t) + \lambda_1|x|^{\chi_1} + \lambda_2|y|^{\chi_2},$$

for each $x \in C_r$, $y \in \mathbb{R}$ where λ_1, λ_2 are negative constants and $\chi_1, \chi_2 > 1$.

Then boundary value problem of Equation (1) has at least one solution.

Proof. The proof is organized into three steps as follows.

Step I. We verify that $\mathcal{T}$ map bounded sets into bounded sets. Let $\max |I_k(u_{t_k-1})| = N$ for $k = 1, 2, ..., p$. Suppose that $(H4)$ holds, we choose a constant:

$$R \geq \max\left\{3\left[\frac{1}{2}p(p+1)N\frac{(T+\alpha-\beta+1)^{1-\beta}}{\Gamma(1-\beta)}+1\right]+\Theta+\widetilde{\Theta}, (3\lambda_1 Y)^{\frac{1}{1-\chi_1}}, (3\lambda_2 Y)^{\frac{1}{1-\chi_2}}\right\}, \quad (21)$$

and define the $\mathcal{P} = \{u \in \mathcal{X} : \|u\| \leq R, R > 0\}$.
For any $u \in \mathcal{P}$, we have:

$$|(\mathcal{T}u)(t)| \leq \sum_{i=1}^{p}|I_i(u_{t_i-1})| + \frac{1}{\Gamma(\alpha)}\sum_{i=1}^{p}\sum_{s=t_{i-1}}^{t_i-1}(t_i-s+\alpha-2)^{\underline{\alpha-1}}\times$$

$$\left[\phi(s)+\lambda_1|u_s|^{\chi_1}+\lambda_2\left|\Delta^{\beta}u(s-\beta+1)\right|^{\chi_2}\right] + \frac{1}{\Gamma(\alpha)}\sum_{s=t_k}^{t-1}(t-s+\alpha-2)^{\underline{\alpha-1}}\times$$

$$\left[\phi(s)+\lambda_1|u_s|^{\chi_1}+\lambda_2\left|\Delta^{\beta}u(s-\beta+1)\right|^{\chi_2}\right]$$

$$\leq \frac{1}{2}p(p+1)N+\Theta+\left(\frac{1}{2}p(p+1)+1\right)\frac{(T+\alpha)^{\underline{\alpha}}}{\Gamma(\alpha)}\left[\lambda_1|u_s|^{\chi_1}+\lambda_2\left|\Delta^{\beta}u(s-\beta+1)\right|^{\chi_2}\right]$$

and

$$|(\Delta^{\beta}\mathcal{T}u)(t-\beta+1)| \leq \frac{1}{\Gamma(-\beta)}\sum_{s=\alpha-1}^{t+1}(t-\beta-s)^{\underline{-\beta-1}}\left[\sum_{i=1}^{p}|I_i(u_{t_i-1})|\right]$$

$$+\frac{1}{\Gamma(\alpha)\Gamma(-\beta)}\sum_{\xi=t_0}^{t+1}\sum_{i=1}^{p}\sum_{s=t_{i-1}}^{t_i-1}(t-\beta-\xi)^{\underline{-\beta-1}}(t_i-s+\alpha-2)^{\underline{\alpha-1}}\times$$

$$\left[\phi(s)+\lambda_1|u_s|^{\chi_1}+\lambda_2\left|\Delta^{\beta}u(s-\beta+1)\right|^{\chi_2}\right]$$

$$+\frac{1}{\Gamma(\alpha)\Gamma(-\beta)}\sum_{\xi=t_0+1}^{t+1}\sum_{s=t_k}^{\xi-1}(t-\beta-\xi)^{\underline{-\beta-1}}(\xi-s+\alpha-2)^{\underline{\alpha-1}}\times$$

$$\left[\phi(s)+\lambda_1|u_s|^{\chi_1}+\lambda_2\left|\Delta^{\beta}u(s-\beta+1)\right|^{\chi_2}\right]$$

$$\leq \frac{1}{2}p(p+1)N\frac{(T+\alpha-\beta+1)^{\underline{-\beta}}}{\Gamma(1-\beta)}+\widetilde{\Theta}+\left(\frac{1}{2}p(p+1)+1\right)\times$$

$$\frac{(T+\alpha)^{\underline{\alpha}}(T+\alpha-\beta+1)^{\underline{-\beta}}}{\Gamma(\alpha)\Gamma(1-\beta)}\left[\lambda_1|u_s|^{\chi_1}+\lambda_2\left|\Delta^{\beta}u(s-\beta+1)\right|^{\chi_2}\right].$$

Hence, we have:

$$\|\mathcal{T}u\|_{\mathcal{X}} \leq \frac{1}{2}p(p+1)N\left[\frac{(T+\alpha-\beta+1)^{\underline{-\beta}}}{\Gamma(1-\beta)}+1\right]+\Theta+\widetilde{\Theta}$$

$$+Y\left[\lambda_1|u_s|^{\chi_1}+\lambda_2\left|\Delta^{\beta}u(s-\beta+1)\right|^{\chi_2}\right]$$

$$\leq \frac{R}{3}+Y\left[\lambda_1|u_s|^{\chi_1}+\lambda_2\left|\Delta^{\beta}u(s-\beta+1)\right|^{\chi_2}\right]$$

$$\leq \frac{R}{3}+\frac{R}{3}+\frac{R}{3} = R. \quad (22)$$

This implies that $\mathcal{T}: \mathcal{P} \to \mathcal{P}$.

For the second case, if $(H5)$ holds, we choose a constant:

$$R \geq \max\left\{3\left[\frac{1}{2}p(p+1)N\frac{(T+\alpha-\beta+1)^{1-\beta}}{\Gamma(1-\beta)}+1\right]+\Theta+\widetilde{\Theta},\left(\frac{1}{3\lambda_1 Y}\right)^{\frac{1}{1-\chi_1}},\left(\frac{1}{3\lambda_2 Y}\right)^{\frac{1}{1-\chi_2}}\right\}. \quad (23)$$

Similarly, we find that:

$$\|\mathcal{T}u(t)\|_{\mathcal{X}} \leq R, \quad (24)$$

which implies that $\mathcal{T}: \mathcal{P} \to \mathcal{P}$.

Step II. It is obvious that the operator $\mathcal{T}$ is continuous on $\mathcal{P}$ since the continuity of F.

Step III. We prove that $\mathcal{T}$ is equicontinuous on $\mathcal{P}$. For any $\epsilon > 0$, there exist positive constants $\delta_1 = \max\{\delta_{11}, \delta_{12}\}$, δ_2 and $\delta_3 = \max\{\delta_{31}, \delta_{32}, \delta_{33}\}$, such that:

(i) for $\tau_1, \tau_2 \in \mathbb{N}_{\alpha-1, T+\alpha}$ and $\tau_1 < \tau_2$

$$|\tau_2^{\underline{\alpha}} - \tau_1^{\underline{\alpha}}| < \frac{\epsilon \Gamma(\alpha+1)}{2M}, \text{ whenever } |\tau_2 - \tau_1| < \delta_{11},$$

$$|(\tau_2 - \beta + 1)^{\underline{-\beta}} - (\tau_1 - \beta + 1)^{\underline{-\beta}}| < \frac{\epsilon \Gamma(\alpha+1)\Gamma(1-\beta)}{2M}, \text{ whenever } |t_2 - t_1| < \delta_{12};$$

(ii) for $\tau_1, \tau_2 \in \mathbb{N}_{\alpha-r-1,\alpha-1}$ and $\tau_1 < \tau_2$

$$|\psi(\tau_2) - \psi(\tau_2)| < \epsilon, \text{ whenever } |\tau_2 - \tau_1| < \delta_2;$$

(iii) for $\tau_1 \in \mathbb{N}_{\alpha-r-1,\alpha-1}$ and $\tau_2 \in \mathbb{N}_{\alpha,T+\alpha}$

$$|\tau_2^{\underline{\alpha}}| < \frac{\epsilon \Gamma(\alpha+1)}{3M}, \text{ whenever } |\tau_2 - \tau_1| < \delta_{31},$$

$$|\psi(\tau_1)| < \frac{\epsilon}{3M}, \text{ whenever } |\tau_2 - \tau_1| < \delta_{32},$$

$$|(\tau_2 - \beta + 1)^{\underline{-\beta}}| < \frac{\epsilon \Gamma(\alpha+1)\Gamma(1-\beta)}{3M}, \text{ whenever } |\tau_2 - \tau_1| < \delta_{33}.$$

Let $M = \max\limits_{t \in \mathbb{N}_{\alpha-1,T+\alpha}}\left|F\left[t, u_t, \Delta^\beta u(t-\beta+1)\right]\right|$. Then, for $u \in \mathcal{P}$ and $\tau_1, \tau_2 \in \mathbb{N}_{\alpha-r-1,T+\alpha}$.

Case 1. If $\tau_1, \tau_2 \in \mathbb{N}_{t_0,t_1} \cup \mathbb{N}_{t_k+1,t_{k+1}}$, $t_{k+1} - t_k \geq 2$, $k = 1, 2, ..., p$, and $\tau_1 < \tau_2$, we obtain:

$$|(\mathcal{T}u)(\tau_2) - (\mathcal{T}u)(\tau_1)| \leq \frac{M}{\Gamma(\alpha)}\left|\sum_{s=t_k}^{\tau_2-1}(\tau_2-s+\alpha-2)^{\underline{\alpha-1}} - \sum_{s=t_k}^{\tau_1-1}(\tau_1-s+\alpha-1)^{\underline{\alpha-1}}\right|$$

$$\leq \frac{M}{\Gamma(\alpha+1)}|\tau_2^{\underline{\alpha}} - \tau_1^{\underline{\alpha}}|$$

$$\leq \frac{\epsilon}{2}, \quad (25)$$

and

$$|(\Delta^\beta \mathcal{T} u)(\tau_2 - \beta + 1) - (\Delta^\beta \mathcal{T} u)(\tau_1 - \beta + 1)|$$
$$\leq \frac{M}{\Gamma(\alpha)\Gamma(-\beta)} \left| \sum_{\xi=t_0}^{\tau_2+1} \sum_{s=t_k}^{\xi-1} (\tau_2 - \beta - \xi)^{\underline{-\beta-1}} (\xi - s + \alpha - 2)^{\underline{\alpha-1}} \right.$$
$$\left. - \sum_{\xi=t_0}^{\tau_1+1} \sum_{s=t_k}^{\xi-1} (\tau_1 - \beta - \xi)^{\underline{-\beta-1}} (\xi - s + \alpha - 2)^{\underline{\alpha-1}} \right|$$
$$\leq \frac{M}{\Gamma(\alpha+1)\Gamma(1-\beta)} \left| (\tau_2 - \beta + 1)^{\underline{-\beta}} - (\tau_1 - \beta + 1)^{\underline{-\beta}} \right|$$
$$\leq \frac{\epsilon}{2}. \tag{26}$$

By Equations (25) and (26), it implies that:

$$\|(\mathcal{T}u)(\tau_2) - (\mathcal{T}u)(\tau_1)\|_{\mathcal{X}} \leq \frac{\epsilon}{2} + \frac{\epsilon}{2} = \epsilon. \tag{27}$$

Case 2. If $\tau_1, \tau_2 \in \mathbb{N}_{\alpha-r-1,\alpha-1}$ and $\tau_1 < \tau_2$, we obtain:

$$|(\mathcal{T}u)(\tau_2) - (\mathcal{T}u)(\tau_1)| = |\psi(\tau_2) - \psi(\tau_1)| < \epsilon, \tag{28}$$

and $\quad |(\Delta^\beta \mathcal{T} u)(\tau_2) - (\Delta^\beta \mathcal{T} u)(\tau_1)| = |(\Delta^\beta \mathcal{T} \psi)(\tau_2) - (\Delta^\beta \mathcal{T} \psi)(\tau_1)| = 0. \tag{29}$

By Equations (28) and (29), it implies that:

$$\|(\mathcal{T}u)(\tau_2) - (\mathcal{T}u)(\tau_1)\|_{\mathcal{X}} < \epsilon + 0 = \epsilon. \tag{30}$$

Case 3. If $\tau_1 \in \mathbb{N}_{\alpha-r-1,\alpha-1}$ and $\tau_2 \in \mathbb{N}_{t_0+1,t_1} \cup \mathbb{N}_{t_k+1,t_{k+1}}$, $t_{k+1} - t_k \geq 2$, $k = 1, 2, ..., p$, we obtain:

$$|(\mathcal{T}u)(\tau_2) - (\mathcal{T}u)(\tau_1)| \leq |(\mathcal{T}u)(\tau_2) - (\mathcal{T}u)(\alpha - 1)| + |(\mathcal{T}u)(\alpha - 1) - (\mathcal{T}u)(\tau_1)|$$
$$\leq (\mathcal{T}u)(\tau_2) + |\psi(\tau_1)|$$
$$\leq \frac{M}{\Gamma(\alpha+1)} |\tau_2^{\underline{\alpha}}| + \|\varphi(\tau_1)\|_{C_r}$$
$$< \frac{\epsilon}{3} + \frac{\epsilon}{3} = \frac{2\epsilon}{3}, \tag{31}$$

and $\quad |(\Delta^\beta \mathcal{T} u)(\tau_2) - (\Delta^\beta \mathcal{T} u)(\tau_1)| \leq \left|(\Delta^\beta \mathcal{T} u)(\tau_2) - 0\right|$
$$\leq \frac{M}{\Gamma(\alpha)\Gamma(-\beta)} \sum_{\xi=t_0}^{\tau_2+1} \sum_{s=t_k}^{\xi-1} (\tau_2 - \beta - \xi)^{\underline{-\beta-1}} (\xi - s + \alpha - 2)^{\underline{\alpha-1}}$$
$$\leq \frac{M}{\Gamma(\alpha+1)\Gamma(1-\beta)} \left| (\tau_2 - \beta + 1)^{\underline{-\beta}} \right|$$
$$\leq \frac{\epsilon}{3}. \tag{32}$$

By Equations (31) and (32), it implies that:

$$\|(\mathcal{T}u)(\tau_2) - (\mathcal{T}u)(\tau_1)\|_{\mathcal{X}} < \frac{2\epsilon}{3} + \frac{\epsilon}{3} = \epsilon. \tag{33}$$

From Steps I to III and the Arzelá–Ascoli theorem, we can conclude that $\mathcal{T} : \mathcal{X} \to \mathcal{X}$ is completely continuous. Therefore, the problem in Equation (1) has at least one solution by Schauder's fixed point theorem. □

5. An Example

Consider the following fractional difference boundary value problem:

$$\Delta_C^{\frac{1}{2}} u(t) = F\left[t - \frac{1}{2}, u_{t-\frac{1}{2}}, \Delta^{\frac{3}{2}} u\left(t - \frac{1}{4}\right)\right], \quad t \in \mathbb{N}_{0,10}, \ t - \frac{1}{2} \neq t_k, \ k = 1, 2, 3$$

$$\Delta u(t_k) = \frac{1}{k + 100} \sin |u(t_k - 1)|, \ t_k = \frac{1}{2} + 2k,$$

$$u\left(-\frac{1}{2}\right) = 0. \tag{34}$$

Here $\alpha = \frac{1}{2}, \beta = \frac{2}{3}, T = 5, p = 3$.

(i) Let $F\left[t, u_t, \Delta^\beta u_{t-\beta+1}\right] = \dfrac{|u_t| + \left|\Delta^{\frac{2}{3}} u\left(t + \frac{1}{3}\right)\right|}{(t+100)^3 \left[1 + |u_t| + \left|\Delta^{\frac{2}{3}} u\left(t + \frac{1}{3}\right)\right|\right]}.$

For $t \in \mathbb{N}_{-\frac{1}{2}, \frac{21}{2}}$, we have:

$$\left|F[t, u_t, \Delta^\beta u] - F[t, v_t, \Delta^\beta v]\right| \leq \frac{8}{7880599}\left[|u_t - v_t| + \left|\Delta^{\frac{2}{3}} u\left(t + \frac{1}{3}\right) - \Delta^{\frac{2}{3}} v\left(t + \frac{1}{3}\right)\right|\right].$$

So, (H1) holds with $\ell = \frac{8}{7880599}$.

For all $u, v \in \mathcal{X}$ and $k = 1, 2, 3$, we have:

$$|I_k(u) - I_k(v)| \leq \frac{1}{101}|u - v|.$$

So, (H2) holds with $\lambda = \frac{1}{101}$.

We can show that (H3) holds with:

$$\left[\frac{p(p+1)}{2}\lambda + \ell\left(\frac{p(p+1)}{2} + 1\right)\frac{(T+\alpha)^\alpha}{\Gamma(\alpha+1)}\right]\left[1 + \frac{(T+\alpha-\beta-1)^{-\beta}}{\Gamma(1-\beta)}\right] \approx 0.06401 < 1.$$

Therefore, by Theorem 1, the boundary value problem of Equation (34) has an unique solution.

(ii) Let $F\left[t, u_t, \Delta^\beta u_{t-\beta+1}\right] = t^2 + \dfrac{e^{-t}}{2(t+1)^3}|u_t|^{\chi_1} + \dfrac{e^{-2(t+1)}}{3}\left|\Delta^{\frac{2}{3}} u_{t+\frac{1}{3}}\right|^{\chi_2}.$

For $t \in \mathbb{N}_{-\frac{1}{2}, \frac{21}{2}}$, we have:

$$\left|F\left[t, u_t, \Delta^\beta u_{t-\beta+1}\right]\right| \leq \left(\frac{21}{2}\right)^2 + \frac{4}{\sqrt{e}}|u_t|^{\chi_1} + \frac{1}{3e}\left|\Delta^{\frac{2}{3}} u\left(t + \frac{1}{3}\right)\right|^{\chi_2}.$$

Thus $|\phi(t)| \leq \frac{441}{4}, \lambda_1 = \frac{4}{\sqrt{e}}, \lambda_2 = \frac{1}{3e}$. We can show that (H4) is satisfied for $0 < \chi_1, \chi_2 < 1$, and (H5) is satisfied for $\chi_1, \chi_2 > 1$. Therefore, by Theorem 2, the boundary value problem in Equation (34) has at least one solution.

6. Conclusions

We established the conditions for the existence and uniqueness of a solution for a nonlinear fractional difference equation with delay and impulses in Equation (1) by using the Banach fixed point theorem, and the conditions of at least one solution by using the Schauder's fixed point theorem. Our problem contained both delay and impulses, which is a new idea.

Author Contributions: Conceptualization, R.O. and S.C.; Formal analysis, R.O. and S.C.; Funding acquisition, S.C.; Investigation, R.O.; Methodology, R.O., S.C., and T.S.; Writing—original draft, R.O., S.C., and T.S.; Writing—review and editing, R.O., S.C., and T.S. All authors have read and agreed to the published version of the manuscript.

Funding: This research was funded by King Mongkut's University of Technology North Bangkok. Contract No. KMUTNB-61-GOV-D-65.

Acknowledgments: This research was supported by Chiang Mai University.

Conflicts of Interest: The authors declare no conflict of interest regarding the publication of this paper.

References

1. Wu, G.C.; Baleanu, D. Discrete fractional logistic map and its chaos. *Nonlinear Dyn.* **2014**, *75*, 283–287. [CrossRef]
2. Wu, G.C.; Baleanu, D. Chaos synchronization of the discrete fractional logistic map. *Signal Process.* **2014**, *102*, 96–99. [CrossRef]
3. Wu, G.C.; Baleanu, D.; Xie, H.P.; Chen F.L. Chaos synchronization of fractional chaotic maps based on stability results. *Physica A* **2016**, *460*, 374–383. [CrossRef]
4. Goodrich, C.S.; Peterson, A.C. *Discrete Fractional Calculus*; Springer: New York, NY, USA, 2015.
5. Abdeljawad, T. On Riemann and Caputo fractional differences, *Comput. Math. Appl.* **2011**, *62*, 1602–1611.
6. Abdeljawad, T. Dual identities in fractional difference calculus within Riemann. *Adv. Differ. Equ.* **2013**, *2013*, 36. [CrossRef]
7. Abdeljawad, T. On delta and nabla Caputo fractional differences and dual identities. *Discrete. Dyn. Nat. Soc.* **2013**, *2013*, 406910. [CrossRef]
8. Atici, F.M.; Eloe, P.W. Two-point boundary value problems for finite fractional difference equations. *J. Differ. Equ. Appl.* **2011**, *17*, 445–456. [CrossRef]
9. Atici, F.M.; Eloe, P.W. A transform method in discrete fractional calculus. *Int. J. Differ. Equ.* **2007**, *2*, 165–176.
10. Atici, F.M.; Eloe, P.W. Initial value problems in discrete fractional calculus. *Proc. Amer. Math. Soc.* **2009**, *137*, 981–989. [CrossRef]
11. Agarwal, R.P.; Leanu, D.; Rezapour, S.; Salehi, S. The existence of solutions for some fractional finite difference equations via sum boundary conditions, *Adv. Differ. Equ.* **2014**, *2014*, 282. [CrossRef]
12. Goodrich, C.S. On discrete sequential fractional boundary value problems. *J. Math. Anal. Appl.* **2012**, *385*, 111–124. [CrossRef]
13. Goodrich, C.S. On a discrete fractional three-point boundary value problem. *J. Differ. Equ. Appl.* **2012**, *18*, 397–415. [CrossRef]
14. Erbe, L.; Goodrich, C.S.; Jia, B.; Peterson, A. Survey of the qualitative properties of fractional difference operators: monotonicity, convexity, and asymptotic behavior of solutions. *Adv. Differ. Equ.* **2016**, *2016*, 43. [CrossRef]
15. Chen, Y.; Tang, X. Three difference between a class of discrete fractional and integer order boundary value problems. *Commun. Nonlinear Sci.* **2014**, *19*, 4057–4067. [CrossRef]
16. Lv, W.; Feng, J. Nonlinear discrete fractional mixed type sum-difference equation boundary value problems in Banach spaces. *Adv. Differ. Equ.* **2014**, *2014*, 184. [CrossRef]
17. Lv, W. Existence of solutions for discrete fractional boundary value problems with a p-Laplacian operator. *Adv. Differ. Equ.* **2012**, *2012*, 163. [CrossRef]
18. Jia, B.; Erbe, L.; Peterson, A. Two monotonicity results for nabla and delta fractional differences. *Arch. Math.* **2015**, *104*, 589–597. [CrossRef]
19. Jia, B.; Erbe, L.; Peterson, A. Convexity for nabla and delta fractional differences. *J. Differ. Equ. Appl.* **2015**, *21*, 360–373.
20. Ferreira, R.A.C. Existence and uniqueness of solution to some discrete fractional boundary value problems of order less than one. *J. Differ. Equ. Appl.* **2013**, *19*, 712–718. [CrossRef]
21. Ferreira, R.A.C.; Goodrich, C.S. Positive solution for a discrete fractional periodic boundary value problem. *Dyn. Contin. Discrete Impuls. Syst. Ser. A Math. Anal.* **2012**, *19*, 545–557.
22. Kang, S.G.; Li, Y.; Chen, H.Q. Positive solutions to boundary value problems of fractional difference equations with nonlocal conditions. *Adv. Differ. Equ.* **2014**, *2014*, 7. [CrossRef]

23. Dong, W.; Xu, J.; Regan, D.O. Solutions for a fractional difference boundary value problem. *Adv. Differ. Equ.* **2013**, *2013*, 319. [CrossRef]
24. Holm, M. Sum and difference compositions in discrete fractional calculus. *Cubo* **2011**, *13*, 153–184. [CrossRef]
25. Sitthiwirattham, T.; Tariboon, J.; Ntouyas S.K. Existence results for fractional difference equations with three-point fractional sum boundary conditions. *Discrete. Dyn. Nat. Soc.* **2013**, *2013*, 104276. [CrossRef]
26. Sitthiwirattham, T.; Tariboon, J.; Ntouyas, S.K. Boundary value problems for fractional difference equations with three-point fractional sum boundary conditions. *Adv. Differ. Equ.* **2013**, *2013*, 296. [CrossRef]
27. Sitthiwirattham, T. Existence and uniqueness of solutions of sequential nonlinear fractional difference equations with three-point fractional sum boundary conditions. *Math. Method. Appl. Sci.* **2015**, *38*, 2809–2815. [CrossRef]
28. Sitthiwirattham, T. Boundary value problem for p−Laplacian Caputo fractional difference equations with fractional sum boundary conditions. *Math. Method. Appl. Sci.* **2016**, *39*, 1522–1534. [CrossRef]
29. Chasreechai, S.; Kiataramkul, C.; Sitthiwirattham, T. On nonlinear fractional sum-difference equations via fractional sum boundary conditions involving different orders. *Math. Probl. Eng.* **2015**, *2015*, 519072 [CrossRef]
30. Reunsumrit, J.; Sitthiwirattham, T. Positive solutions of three-point fractional sum boundary value problem for Caputo fractional difference equations via an argument with a shift. *Positivity* **2016**, *20*, 861–876. [CrossRef]
31. Reunsumrit, J.; Sitthiwirattham, T. On positive solutions to fractional sum boundary value problems for nonlinear fractional difference equations. *Math. Method. Appl. Sci.* **2016**, *39*, 2737–2751. [CrossRef]
32. Reunsumrit, R.; Sitthiwirattham, T. a New class of four-point fractional sum boundary value problems for nonlinear sequential fractional difference equations involving shift operators. *Kragujevac J. Math.* **2018**, *42*, 371–387. [CrossRef]
33. Chasreechai, S.; Sitthiwirattham, T. Existence results of initial value problems for hybrid fractional sum-difference equations. *Discrete Dyn. Nat. Soc.* **2018**, *2018*, 5268528. [CrossRef]
34. Chasreechai, S.; Sitthiwirattham, T. On separate fractional sum-difference boundary value problems with n-point fractional sum-difference boundary conditions via arbitrary different fractional orders. *Mathematics* **2019**, *2019*, 471. [CrossRef]
35. Kunnawuttipreechachan, E.; Promsakon, C.; Sitthiwirattham, T. Nonlocal fractional sum boundary value problem for a coupled system of fractional sum-difference equations. *Dyn. Syst. Appl.* **2019**, *28*, 73–92. [CrossRef]
36. Promsakon, C.; Chasreechai, S.; Sitthiwirattham, T. Positive solution to a coupled system of singular fractional difference equations with fractional sum boundary value conditions. *Adv. Differ. Equ.* **2019**, *2019*, 218. [CrossRef]
37. Soontharanon, J.; Chasreechai, S.; Sitthiwirattham, T. On a coupled system of fractional difference equations with nonlocal fractional sum boundary value conditions on the discrete half-line. *Mathematics* **2019**, *2019*, 256. [CrossRef]
38. Kaewwisetkul, B.; Sitthiwirattham, T. On nonlocal fractional sum-difference boundary value problems for Caputo fractional functional difference equations with delay. *Adv. Differ. Equ.* **2017**, *2017*, 219. [CrossRef]
39. Wu, G.C.; Baleanu, D.; Zeng, S.D. Finite-time stability of discrete fractional delay systems: Gronwall inequality and stability criterion. *Commun. Nonlinear Sci. Numer. Simulat.* **2018**, *57*, 299–308. [CrossRef]
40. Alzabut, J.; Abdeljawad, T.; Baleanu, D. Nonlinear delay fractional difference equations with applications on discrete fractional Lotka—Volterra competition model. *J. Comput. Anal. Appl.* **2018**, *25*, 889–898.
41. Alzabut, J.; Abdeljawad, T. A generalized discrete fractional Gronwall inequality and its application on the uniqueness of solutions for nonlinear delay fractional difference system. *Appl. Anal. Discr. Math.* **2018**, *12*, 36–48. [CrossRef]
42. Luo, D.; Luo, Z. Uniqueness and Novel Finite-Time Stability of Solutions for a Class of Nonlinear Fractional Delay Difference Systems. *Discrete Dyn. Nat. Soc.* **2018**, *2018*, 8476285. [CrossRef]
43. Wu, G.C.; Baleanu, D.; Huang, L.L. Novel Mittag-Leffler stability of linear fractional delay difference equations with impulse. *Appl. Math. Lett.* **2018**, *82*, 71–78. [CrossRef]

44. Griffel, D.H. *Applied Functional Analysis*; Ellis Horwood Publishers: Chichester, UK, 1981.
45. Guo, D.; Lakshmikantham, V. *Nonlinear Problems in Abstract Cone*; Academic Press: Orlando, FL, USA, 1988.

© 2020 by the authors. Licensee MDPI, Basel, Switzerland. This article is an open access article distributed under the terms and conditions of the Creative Commons Attribution (CC BY) license (http://creativecommons.org/licenses/by/4.0/).

Article

A Collocation Approach for Solving Time-Fractional Stochastic Heat Equation Driven by an Additive Noise

Afshin Babaei [1], Hossein Jafari [2,3,4,*] and S. Banihashemi [1]

1 Department of Mathematics, University of Mazandaran, Babolsar 4741613534, Iran; babaei@umz.ac.ir (A.B.); s.banihashemi@stu.umz.ac.ir (S.B.)
2 Institute of Research and Development, Duy Tan University, Da Nang 550000, Vietnam
3 Faculty of Natural Sciences, Duy Tan University, Da Nang 550000, Vietnam
4 Department of Medical Research, China Medical University Hospital, China Medical University, Taichung 110122, Taiwan
* Correspondence: jafarh@unisa.ac.za

Received: 27 April 2020; Accepted: 21 May 2020; Published: 1 June 2020

Abstract: A spectral collocation approach is constructed to solve a class of time-fractional stochastic heat equations (TFSHEs) driven by Brownian motion. Stochastic differential equations with additive noise have an important role in explaining some symmetry phenomena such as symmetry breaking in molecular vibrations. Finding the exact solution of such equations is difficult in many cases. Thus, a collocation method based on sixth-kind Chebyshev polynomials (SKCPs) is introduced to assess their numerical solutions. This collocation approach reduces the considered problem to a system of linear algebraic equations. The convergence and error analysis of the suggested scheme are investigated. In the end, numerical results and the order of convergence are evaluated for some numerical test problems to illustrate the efficiency and robustness of the presented method.

Keywords: fractional calculus; stochastic heat equation; additive noise; chebyshev polynomials of sixth kind; error estimate

1. Introduction

Many models in physics, chemistry, and engineering reveal stochastic effects and are introduced as stochastic partial differential equations (SPDEs) [1,2]. Some phenomena in various fields such as population dynamics [3], motions of ions in crystals [4], optimal pricing in economics [5] and thermal noise [6] show stochastic behaviors. Fractional stochastic partial differential equation (FSPDE) is an example of these equations that have attracted more attention recently.

In recent decades, investigations have shown that fractional calculus provides some new ways for a better understanding of behaviors of real-world phenomena. Fractional-order operators give helpful tools for modeling inherited memory characteristics of real applications. Scientists proposed models for numerous phenomena in engineering, fluid mechanics, physics [7–11], finance [12,13], geomagnetic [14] and hydrology [15] based on fractional differential and integral equations. Non-Markovian anomalous diffusion in materials with memory, such as, viscoelastic substances is an example of these applications [16], in which the mean square displacement of particles grows faster or slower than in the case of normal diffusion.

In many applications, it is more realistic to represent the mathematical model of the problem in a non-deterministic state. In other words, some kinds of randomness and uncertainty are considered in the mathematical formulation of the problem. Hence, stochastic functional equations have arisen in many situations and numerous problems in different fields of science are modeled as fractional stochastic differential or integral equations [17–19]. Many theoretical investigations on the fractional stochastic differential equations have been made by researchers in the literature. Liu et al. studied some properties of fractional stochastic heat equations [20]. Ralchenko and Shevchenko [21] surveyed the existence and uniqueness of mild solution for a special type of stochastic heat equations of fractional order. Roozbahani et al. [22] proved the unique solvability of a class of SPDEs. Moghaddam et al. [23] proved the existence and uniqueness of solution for some delay stochastic differential equations of fractional order. Moreover, Mishura et al. [24] investigated mild and weak solutions for a SPDE with second order elliptic operator in divergence form. Since the exact solutions of these equations are scarcely known, researchers have examined several numerical algorithms to solve them. Finite difference schemes [25,26], finite element approaches [27–29], wavelets Galerkin method [30], B-spline collocation method [31,32], hat function operational matrix method [33], mean-square dissipative method [34] and operational matrix of Chebyshev wavelets [35] are a number of these schemes.

In the present work, we consider the following TFSHE

$$D_{0,t}^{\alpha} u(x,t) = \left(\mu + \vartheta \dot{B}(t)\right) u_{xx}(x,t) + \lambda u_x(x,t) + f(x,t), \tag{1}$$

where $(x,t) \in \mathcal{L} \times \mathcal{I}$, with the boundary and initial conditions

$$u(x,t) = \varphi(x,t), \quad x \in \partial \mathcal{L}, \ t \in \mathcal{I}, \tag{2}$$
$$u(x,0) = \eta(x), \quad x \in \mathcal{L}, \tag{3}$$

where $\alpha \in (0,1)$, μ, ϑ and λ are real constants, $\mathcal{I} := [0,T]$, $\mathcal{L} := [0,l]$ and $\partial \mathcal{L}$ is the boundary of $\mathcal{L}$. Also, $\dot{B}(t) := \frac{dB(t)}{dt}$ denotes a time white noise where $B(t)$, $t \in \mathcal{I}$ is the Brownian motion adapted to a filtration $F_B = \{F_t\}_{t \in \mathcal{I}}$ in a probability space $(\Omega_B, F_B, \mathbb{P}_B)$ [36]. Moreover, the source term $f(x,t)$, $\varphi(x,t)$ and $\eta(x)$ are some stochastic processes defined on $(\Omega_B, F_B, \mathbb{P}_B)$ and $u(x,t)$ is an unknown stochastic function to be found. Moreover, the operator $D_{0,t}^{\alpha}[\cdot]$ denotes Caputo fractional derivative defined as:

$$D_{0,t}^{\alpha} u(x,t) = \frac{1}{\Gamma(1-\alpha)} \int_0^t \frac{1}{(t-\xi)^{\alpha}} \frac{\partial u}{\partial \xi}(x,\xi) \, d\xi, \quad \alpha \in (0,1), \tag{4}$$

and $\Gamma(\cdot)$ represents the Gamma function.

Equation (1) is a FSPDE driven by additive noise that takes into account both memory and environmental noise effects. Many physical and engineering models are built based on these types of stochastic equations. Fractional stochastic heat equations [20,37–39], stochastic Burgers equation [40] and stochastic coupled fractional Ginzburg-Landau equation [41] are some examples of these applications. The problem (1)–(3) has been considered in [30], in the case $\alpha = 1$. The authors have proposed a wavelet Galerkin method to find the solution to this equation. When $\vartheta = 0$, Equation (1) reduces to an advection-dispersion equation of fractional order describing the transport of passive tracers in a porous medium in groundwater hydrology [42].

Many numerical schemes with Chebyshev polynomials basis functions are established in literature to solve various types of problems. Masjed-Jamei in [43] introduced a class of symmetric orthogonal polynomials. The six various types of Chebyshev polynomials are special cases of this basic class. To our experience, the approaches based on the SKCPs expansions result very accurate numerical estimations. Hence, we motivated to employ this kind of Chebyshev polynomials for solving TFSHEs. Recently, a few authors applied the SKCPs to solve some types of differential equations [44–46].

The structure of this work is organized as follows. In Section 2, the basic concepts of the SKCPs theory are described. In Section 3, the collocation scheme based on the SKCPs is applied. The convergence of the numerical procedure is considered in Section 4. The accuracy of the proposed approach is analyzed in Section 5 by three numerical test problems. In the end, the main concluding remarks are presented in Section 6.

2. The Shifted SKCPs and Their Properties

In this section, some necessary preliminaries and relevant properties of the shifted SKCPs utilized in the next sections, are reviewed.

Definition 1. *The shifted SKCPs on $[0, l]$ are defined by*

$$\mathcal{J}_m(x) = \hat{\mathcal{J}}_m((2/l)x - 1), \qquad m = 0, 1, 2, \ldots,$$

where ([43])

$$\hat{\mathcal{J}}_m(x) = \prod_{i=0}^{\lfloor \frac{m}{2} \rfloor - 1} \frac{2i + (-1)^{m+1} + 4}{-5 - (2i + 2\lfloor \frac{m}{2} \rfloor + (-1)^{m+1})} \mathcal{E}_m(x), \qquad (5)$$

and

$$\mathcal{E}_m(x) = \sum_{\tau=0}^{\lfloor \frac{m}{2} \rfloor} \prod_{\kappa=0}^{\lfloor \frac{m}{2} \rfloor - (\tau+1)} \left(\frac{(-1)^m - 2(\kappa + \lfloor \frac{m}{2} \rfloor) - 5}{(-1)^{m+1} + 2(\kappa + 2)} \right) \frac{(\lfloor \frac{m}{2} \rfloor)!}{\tau!(\tau - \lfloor \frac{m}{2} \rfloor)!} x^{m-2\tau}. \qquad (6)$$

The explicit form of shifted SKCPs as follows: [45]

$$\mathcal{J}_m(x) = \sum_{r=0}^{m} \bar{\theta}_{r,m}(x/l)^r, \qquad (7)$$

where

$$\bar{\theta}_{r,m} = \begin{cases} \dfrac{2^{2r-m}}{(2r+1)!} \sum_{i=\lfloor \frac{r+1}{2} \rfloor}^{\frac{m}{2}} \dfrac{(-1)^{\frac{m}{2}+i+r}(2i+r+1)!}{(2i-r)!}, & m \text{ even}, \\[2mm] \dfrac{2^{2r-m+1}}{(m+1)(2r+1)!} \sum_{i=\lfloor \frac{r}{2} \rfloor}^{\frac{m-1}{2}} \dfrac{(-1)^{\frac{m+1}{2}+i+r}(i+1)(2i+r+2)!}{(2i-r+1)!}, & m \text{ odd}. \end{cases}$$

Theorem 1. *([46]) Suppose $L^2_W(\tilde{\Lambda})$ is the square integrable function space according to the weight $W(x,t) = (2x-1)^2(2t-1)^2\sqrt{x-x^2}\sqrt{t-t^2}$. Let $g(x,t) \in L^2_W(\tilde{\Lambda})$ is considered with $\left\|\frac{\partial^6 g(x,t)}{\partial x^3 \partial t^3}\right\|_2 \leq \varsigma$ for some constant $\varsigma > 0$, satisfies the expansion $g(x,t) = \sum_{i=0}^{\infty}\sum_{j=0}^{\infty} c_{i,j}\mathcal{J}_i(x)\mathcal{J}_j(t)$. If*

$$G_{N,M}(x,t) = \sum_{i=0}^{N}\sum_{j=0}^{M} c_{i,j}\mathcal{J}_i(x)\mathcal{J}_j(t), \qquad (8)$$

is an approximation of $g(x,t)$, then

$$|g(x,t) - G_{N,M}(x,t)| < \frac{\varsigma}{2^{N+M}},$$

$$\left|\frac{\partial g}{\partial x}(x,t) - \frac{\partial G_{N,M}}{\partial x}(x,t)\right| < \xi \frac{N}{2^{N+M-2}},$$

$$\left|\frac{\partial^2 g}{\partial x^2}(x,t) - \frac{\partial^2 G_{N,M}}{\partial x^2}(x,t)\right| < \varrho \frac{N^3}{2^{N+M-8}},$$

where ξ and ϱ are two positive constants.

3. The SKCPs-Collocation Approach

In the following, we describe a numerical technique to solve problem (1)–(3). For this reason, we consider the numerical solution of (1) as follows

$$u(x,t) \simeq U_{N,M}(x,t) = \sum_{i=0}^{N}\sum_{j=0}^{M} \delta_{i,j}\mathcal{J}_i(x)\bar{\mathcal{J}}_j(t) = \mathbf{J}(x)^T \mathbf{C}\bar{\mathbf{J}}(t), \qquad (9)$$

where

$$\mathbf{J}(x) = [\mathcal{J}_0(x), \ldots, \mathcal{J}_i(x), \ldots, \mathcal{J}_N(x)]^T, \qquad (10)$$

$$\bar{\mathbf{J}}(t) = [\bar{\mathcal{J}}_0(t), \ldots, \bar{\mathcal{J}}_j(t), \ldots, \bar{\mathcal{J}}_M(t)]^T, \qquad (11)$$

in which $\mathcal{J}_i(x) = \hat{\mathcal{J}}_i((2/l)x - 1)$ on the interval $\mathcal{L}$ and $\bar{\mathcal{J}}_j(t) = \hat{\mathcal{J}}_i((2/T)t - 1)$ on the interval $\mathcal{I}$. Moreover

$$\mathbf{C} = \begin{pmatrix} \delta_{0,0} & \cdots & \delta_{0,M} \\ \vdots & \ddots & \vdots \\ \delta_{N,0} & \cdots & \delta_{N,M} \end{pmatrix}_{(N+1)\times(M+1)},$$

is an unknown coefficients matrix.

Theorem 2. *Let $\bar{\mathbf{J}}(t)$ is the shifted SKCPs vector as (11), then*

$$D_{0,t}^{\alpha} \bar{\mathbf{J}}(t) = \Phi^{\alpha}(t), \qquad (12)$$

where $\Phi^{\alpha}(t)$ is Caputo's fractional derivative of the vector $\bar{\mathbf{J}}(t)$ and is defined as

$$\Phi^{\alpha}(t) = \left[0, \sum_{r=1}^{1} \psi_{r,1}^{\alpha}(t), \ldots, \sum_{r=1}^{j} \psi_{r,j}^{\alpha}(t), \ldots, \sum_{r=1}^{M} \psi_{r,M}^{\alpha}(t) \right]^T, \qquad (13)$$

where

$$\psi_{r,j}^{\alpha}(t) = \frac{\Gamma(r+1)}{T^r \Gamma(r+1-\alpha)} \bar{\theta}_{r,j} \, t^{r-\alpha}.$$

Proof. Due to the analytic form (7), we have

$$D_{0,t}^{\alpha} \bar{\mathcal{J}}_0(t) = \bar{\theta}_{0,0} D_{0,t}^{\alpha}(1) = 0, \qquad (14)$$

Also, we know that [7]

$$D_{0,t}^{\alpha} t^r = \frac{\Gamma(r+1)}{\Gamma(r+1-\alpha)} t^{r-\alpha}. \qquad (15)$$

for $r \geq 1$. So, for $j = 1, \ldots, M$, we get

$$D_{0,t}^{\alpha} \bar{\mathcal{J}}_j(t) = \sum_{r=0}^{j} \bar{\theta}_{r,j} D_{0,t}^{\alpha}(t/T)^r = \sum_{r=1}^{j} \psi_{r,j}^{\alpha}(t)$$

in which $\psi_{r,j}^{\alpha}(t) = \frac{\Gamma(r+1)}{T^r \Gamma(r+1-\alpha)} \bar{\theta}_{r,j} t^{r-\alpha}$. □

According to Equations (1) and (9) and by applying Theorem 2, we have

$$\mathbf{J}(x)^T \mathbf{C} \Phi^{\alpha}(t) = (\mu + \vartheta \dot{B}(t)) \mathbf{J}_{xx}(x)^T \mathbf{C} \bar{\mathbf{J}}(t) + \lambda \mathbf{J}_x(x)^T \mathbf{C} \bar{\mathbf{J}}(t) + f(x,t), \qquad (16)$$

where

$$\mathbf{J}_x(x) = [\mathcal{J}_0'(x), \cdots, \mathcal{J}_i'(x), \cdots, \mathcal{J}_N'(x)]^T,$$

$$\mathbf{J}_{xx}(x) = [\mathcal{J}_0''(x), \cdots, \mathcal{J}_i''(x), \cdots, \mathcal{J}_N''(x)]^T,$$

and from the conditions (2) and (3) and Equation (9), we have

$$\mathbf{J}(0)^T \mathbf{C}\bar{\mathbf{J}}(t) = \varphi(0,t), \tag{17}$$

$$\mathbf{J}(l)^T \mathbf{C}\bar{\mathbf{J}}(t) = \varphi(l,t), \tag{18}$$

$$\mathbf{J}(x)^T \mathbf{C}\bar{\mathbf{J}}(0) = \eta(x). \tag{19}$$

Let $x_0 = 0$, $x_N = l$, and $x_1, \ldots, x_{N-1}$, are the roots of $\mathcal{J}_{N-1}(x)$. Also, suppose t_j, $j = 1, \ldots, M$, are roots of $\tilde{\mathcal{J}}_M(t)$. By considering these collocation nodes, we define

$$\Lambda = [\mathbf{J}(x_1), \ldots, \mathbf{J}(x_i), \ldots, \mathbf{J}(x_{N-1})]^T, \tag{20}$$

$$\Lambda_x = [\mathbf{J}_x(x_1), \ldots, \mathbf{J}_x(x_i), \ldots, \mathbf{J}_x(x_{N-1})]^T, \tag{21}$$

$$\Lambda_{xx} = [\mathbf{J}_{xx}(x_1), \ldots, \mathbf{J}_{xx}(x_i), \ldots, \mathbf{J}_{xx}(x_{N-1})]^T, \tag{22}$$

where the matrices Λ, Λ_x and Λ_{xx} are of the order $(N-1) \times (N+1)$ and

$$\Psi = [\bar{\mathbf{J}}(t_1), \ldots, \bar{\mathbf{J}}(t_j), \ldots, \bar{\mathbf{J}}(t_M)]_{(M+1) \times M}, \tag{23}$$

$$\Psi^\alpha = [\Phi^\alpha(t_1), \ldots, \Phi^\alpha(t_j), \ldots, \Phi^\alpha(t_M)]_{(M+1) \times M}. \tag{24}$$

By evaluating (16) at $(N-1) \times M$ collocation points (x_i, t_j) for $i = 1, \ldots, N-1$ and $j = 1, \ldots, M$, we have

$$\Lambda \mathbf{C} \Psi^\alpha = \Lambda_{xx} \mathbf{C} \Psi \mathcal{B} + \lambda \Lambda_x \mathbf{C} \Psi + \mathcal{F}, \tag{25}$$

where

$$\mathcal{B} = diag\left(\mu + \vartheta b_1, \cdots, \mu + \vartheta b_j, \cdots, \mu + \vartheta b_M\right),$$

in which $b_j = B(t_j) - B(t_{j-1})$, $t_0 = 0$ and

$$\mathcal{F} = \left[f_{i,j}\right]_{(N-1) \times M}, \quad f_{i,j} = f(x_i, t_j), \; i = 1, \ldots, N-1, \; j = 1, \ldots, M.$$

Also, by evaluating (17) and (18) at collocation points t_j and (19) at collocation points x_i, we get

$$\mathbf{J}(0)^T \mathbf{C} \Psi = Y_0, \tag{26}$$

$$\mathbf{J}(l)^T \mathbf{C} \Psi = Y_l, \tag{27}$$

$$\bar{\Lambda} \mathbf{C} \bar{\mathbf{J}}(0) = \tilde{Y}, \tag{28}$$

where

$$Y_0 = [\varphi(0, t_1), \ldots, \varphi(0, t_j), \ldots, \varphi(0, t_M)]^T, \quad Y_l = [\varphi(l, t_1), \ldots, \varphi(l, t_j), \ldots, \varphi(l, t_M)]^T,$$

$$\bar{\Lambda} = [\mathbf{J}(x_0), \ldots, \mathbf{J}(x_i), \ldots, \mathbf{J}(x_N)]^T, \quad \tilde{Y} = [\eta(x_0), \ldots, \eta(x_i), \ldots, \eta(x_N)]^T.$$

Using the Kronecker product, Equation (25) transforms to

$$\mathcal{A}\mathcal{X} = \mathcal{T}_{\text{vec}}, \tag{29}$$

where

$$\mathcal{A} = \Psi^{\alpha T} \otimes \Lambda - (\Psi \mathcal{B})^T \otimes \Lambda_{xx} - \lambda \Psi^T \otimes \Lambda_x,$$

and $\mathcal{X} = vec(\mathbf{C})$, $\mathcal{T}_{\text{vec}} = vec(\mathcal{F})$. Also, Equations (26)–(28) are equivalent to

$$\bar{E}\mathcal{X} = \tilde{Y}, \quad E_0 \mathcal{X} = Y_0, \quad E_l \mathcal{X} = Y_l, \tag{30}$$

where
$$\bar{\mathbf{E}} = \bar{\mathbf{J}}(0)^T \otimes \bar{\Lambda}, \qquad \mathbf{E}_0 = \Psi^T \otimes \mathbf{J}(0)^T, \qquad \mathbf{E}_l = \Psi^T \otimes \mathbf{J}(l)^T.$$

Thus, from Equations (29) and (30), we obtain a system of linear equations $\mathbf{A}\mathcal{X} = \mathbf{B}$ in which
$$\mathbf{A} = \left[\mathcal{A}^T, \bar{\mathbf{E}}^T, \mathbf{E}_0^T, \mathbf{E}_l^T\right]^T, \qquad \mathbf{B} = \left[\mathcal{T}_{\text{vec}}^T, \tilde{\mathbf{Y}}^T, \mathbf{Y}_0^T, \mathbf{Y}_l^T\right]^T.$$

Solving this system leads to an estimation $\mathtt{U}_{N,M}(x,t)$ for the solution of (1)–(3), in the form (9).

4. Convergence Analysis

In the following, we examine the convergence of the approximate solution expressed in the form (9) for the problem (1)–(3).

Theorem 3. *Let $\mathtt{U}_{N,M}(x,t)$ is the approximate solution obtained by the procedure presented in Section 3 and $u(x,t)$ is the exact solution of (1)–(3). Consider the residual error $\mathcal{R}_{N,M}(x,t)$ of this numerical solution. Then, $\mathbb{E}\|\mathcal{R}_{N,M}(x,t)\|_\infty$ tends to zero, when $N \to \infty$ and $M \to \infty$.*

Proof. Suppose $\mathtt{U}_{N,M}(\mathrm{x},\mathrm{t})$, for $(\mathrm{x},\mathrm{t}) \in \mathcal{L} \times \mathcal{I}$, satisfies the equation

$$D_{0,t}^\alpha \mathtt{U}_{N,M}(\mathrm{x},\mathrm{t}) = (\mu + \vartheta \dot{B}(\mathrm{t})) \frac{\partial^2 \mathtt{U}_{N,M}}{\partial x^2}(x,t)$$
$$+ \lambda \frac{\partial \mathtt{U}_{N,M}}{\partial x}(\mathrm{x},\mathrm{t}) + f(\mathrm{x},\mathrm{t}) + \mathcal{R}_{N,M}(\mathrm{x},\mathrm{t}), \tag{31}$$

where $\mathcal{R}_{N,M}(\mathrm{x},\mathrm{t})$ is the residual function. Now, from Equations (1) and (31), we get

$$\mathbb{E}\|\mathcal{R}_{N,M}(\mathrm{x},\mathrm{t})\|_\infty \leq \mathbb{E}\left\|D_{0,t}^\alpha\left(u(\mathrm{x},\mathrm{t}) - \mathtt{U}_{N,M}(\mathrm{x},\mathrm{t})\right)\right\|_\infty$$
$$+ \mathbb{E}\left(\|\mu + \vartheta \dot{B}(\mathrm{t})\|_\infty \left\|u_{\mathrm{xx}}(\mathrm{x},\mathrm{t}) - \frac{\partial^2 \mathtt{U}_{N,M}}{\partial x^2}(\mathrm{x},\mathrm{t})\right\|_\infty\right)$$
$$+ |\lambda|\,\mathbb{E}\left\|u_{\mathrm{x}}(\mathrm{x},\mathrm{t}) - \frac{\partial \mathtt{U}_{N,M}}{\partial x}(\mathrm{x},\mathrm{t})\right\|_\infty. \tag{32}$$

By using Theorem 1, we have

$$\left\|u_t(\mathrm{x},\mathrm{t}) - \frac{\partial \mathtt{U}_{N,M}}{\partial t}(\mathrm{x},\mathrm{t})\right\|_\infty = \sup_{(\mathrm{x},\mathrm{t})\in\mathcal{L}\times\mathcal{I}} \left|u_t(\mathrm{x},\mathrm{t}) - \frac{\partial \mathtt{U}_{N,M}}{\partial t}(\mathrm{x},\mathrm{t})\right|$$
$$< \frac{\theta_1 M}{2^{N+M-2}},$$

where θ_1 is a positive constant, thus

$$\mathbb{E}\left\|D_{0,t}^\alpha\left(u(\mathrm{x},\mathrm{t}) - \mathtt{U}_{N,M}(\mathrm{x},\mathrm{t})\right)\right\|_\infty \leq \mathbb{E}\left(\int_0^t \frac{\|(t-\tau)^{-\alpha}\|_\infty}{\Gamma(1-\alpha)}\left\|u_\tau(\mathrm{x},\tau) - \frac{\partial \mathtt{U}_{N,M}}{\partial \tau}(\mathrm{x},\tau)\right\|_\infty d\tau\right)$$
$$< \frac{\theta_1 M}{\Gamma(1-\alpha)2^{N+M-2}} \mathbb{E}\left(\int_0^t \|(t-\tau)^{-\alpha}\|_\infty d\tau\right).$$

Since $0 < \tau < t \leq T$, hence, we get

$$\mathbb{E}\left\|D_{0,t}^\alpha\left(u(\mathrm{x},\mathrm{t}) - \mathtt{U}_{N,M}(\mathrm{x},\mathrm{t})\right)\right\|_\infty < \frac{\theta_1 T^{1-\alpha} M}{\Gamma(1-\alpha)2^{N+M-2}}. \tag{33}$$

Also, from Theorem 1, we have

$$\left\| u_{xx}(x,t) - \frac{\partial^2 U_{N,M}}{\partial x^2}(x,t) \right\|_\infty = \sup_{(x,t)\in \mathcal{L}\times\mathcal{I}} \left| u_{xx}(x,t) - \frac{\partial^2 U_{N,M}}{\partial x^2}(x,t) \right|$$
$$< \frac{\theta_2 N^3}{2^{N+M-8}}, \qquad (34)$$

$$\left\| u_x(x,t) - \frac{\partial U_{N,M}}{\partial x}(x,t) \right\|_\infty = \sup_{(x,t)\in \mathcal{L}\times\mathcal{I}} \left| u_x(x,t) - \frac{\partial U_{N,M}}{\partial x}(x,t) \right|$$
$$< \frac{\theta_3 N}{2^{N+M-2}}, \qquad (35)$$

where θ_2 and θ_3 are positive constants. Let $\bar{\gamma} = \|\dot{B}(t)\|_\infty$, then, from the relations (32)–(35), it can be concluded that

$$\mathbb{E}\|\mathcal{R}_{N,M}(x,t)\|_\infty < \frac{\theta_1 T^{1-\alpha} M}{\Gamma(1-\alpha) 2^{N+M-2}} + (|\mu| + \bar{\gamma}|\vartheta|)\frac{\theta_2 N^3}{2^{N+M-8}} + |\lambda|\frac{\theta_3 N}{2^{N+M-2}}$$
$$< \frac{\theta_1 T^{1-\alpha} M}{\Gamma(1-\alpha) 2^{N+M-8}} + (|\mu| + \bar{\gamma}|\vartheta|)\frac{\theta_2 N^3}{2^{N+M-8}} + |\lambda|\frac{\theta_3 N^3}{2^{N+M-8}}$$
$$< \hat{\theta}\frac{M + 2N^3}{2^{N+M-8}}, \qquad (36)$$

where

$$\hat{\theta} = \max\{\frac{\theta_1 T^{1-\alpha}}{\Gamma(1-\alpha)}, (|\mu| + \bar{\gamma}|\vartheta|)\theta_2, |\lambda|\theta_3\}.$$

Moreover, for $x \in \partial\mathcal{L}$ and $t \in \mathcal{I}$, $U_{N,M}(x,t)$ satisfies the following equation

$$\mathbb{E}\|\mathcal{R}_{N,M}(x,t)\|_\infty = \mathbb{E}\|\varphi(x,t) - U_{N,M}(x,t)\|_\infty$$
$$= \mathbb{E}\|u(x,t) - U_{N,M}(x,t)\|_\infty$$
$$= \sup_{(x,t)\in\partial\mathcal{L}\times\mathcal{I}} \left| u(x,t) - U_{N,M}(x,t) \right| < \frac{\theta_4}{2^{N+M}}, \qquad (37)$$

and for $x \in \mathcal{L}$, we have

$$\mathbb{E}\|\mathcal{R}_{N,M}(x,0)\|_\infty = \mathbb{E}\|\eta(x) - U_{N,M}(x,0)\|_\infty$$
$$= \mathbb{E}\|u(x,0) - U_{N,M}(x,0)\|_\infty$$
$$= \sup_{x\in\mathcal{L}} \left| u(x,0) - U_{N,M}(x,0) \right| < \frac{\theta_5}{2^{N+M}}. \qquad (38)$$

Therefore, from Equations (36)–(38), we can see that $\mathbb{E}\|\mathcal{R}_{N,M}(x,t)\|_\infty$ tends to zero, when $N, M \to \infty$. □

5. Applications and Results

We assess the applicability of our proposed approach to solve some stochastic heat equations of fractional order.

To simulate the Brownian motion $B(t)$, we employ the approach described in [36]. Consider a discretization of $B(t)$. We set $t_0 = 0$ and let $t_j, j = 1, \ldots, M$, are the considered collocation points, where $t_i < t_j$ for $i < j$. Also, let $B_j = B(t_j)$ and

$$\Delta_j = t_j - t_{j-1}, \quad j = 1, \ldots, M. \qquad (39)$$

From the definition of Brownian motion $B(t)$ on $(\Omega_B, \mathcal{F}_B, \mathbb{P}_B)$, we know that $\mathscr{B}(0) = 0$ with the probability 1. $\mathscr{B}(\tau) - \mathscr{B}(r) \sim \sqrt{\tau - r}\mathcal{N}(0,1)$, for $0 \leq r < \tau \leq T$, where $\mathcal{N}(0,1)$ is a normally distributed random variable with zero mean and unit variance. Also, $\mathscr{B}(\tau_2) - \mathscr{B}(\tau_1)$ and $\mathscr{B}(\nu_2) - \mathscr{B}(\nu_1)$ are independent for $0 \leq \tau_1 < \tau_2 < \nu_1 < \nu_2 \leq T$. Thus, we let $B_0 = t_0$ with the probability 1, and

$$B_j = B_{j-1} + dB_j, \qquad j = 1, \ldots, M, \tag{40}$$

where each dB_j is an independent random variable of the form $\sqrt{\Delta_j}\mathcal{N}(0,1)$. Throughout the section, unless stated otherwise, we assume that $T = 1, l = 1$ and $N = M$. Also, we evaluate the numerical solution $u(x,t)$ along $\bar{P}$ discretized paths and finally, the average of the results over these paths is considered.

The L_∞-norm error is evaluated using the following definition:

$$\|E_N\|_\infty = \max_{1 \leq i,j \leq N} |u(\xi_i, \tau_j) - \mathcal{U}_N(\xi_i, \tau_j)|, \tag{41}$$

where $\mathcal{U}_N(\xi_i, \tau_j)$ and $u(x,t)$, are computed by the exact and numerical solutions defined in (9) at the collocation points $x = \xi_i$ and $t = \tau_j$, respectively. The convergence order is defined by the following formula:

$$\text{CO} = \log_{\frac{N_1}{N_2}} \frac{\|E_{N_1}\|_\infty}{\|E_{N_2}\|_\infty}, \tag{42}$$

where $\|E_{N_i}\|_\infty$ denotes the L_∞-norm error for N_i ($i = 1, 2$) collocation points. The numerical computations are performed on a personal computer using a 1.70 GHz processor and the codes are written in Matlab software.

Example 1. *Consider the time-fractional stochastic equation*

$$D_{0,t}^\alpha u(x,t) = \dot{B}(t)u_{xx}(x,t) + u_x(x,t) + f(x,t),$$

subject to the conditions:

$$u(x,0) = 0,$$
$$u(0,t) = 0, \quad u(1,t) = \alpha \exp(1)t^2,$$

where $\alpha \in (0,1)$, $B(t)$ is a Brownian motion and

$$f(x,t) = \frac{\alpha\Gamma(3)}{\Gamma(3-\alpha)}t^{2-\alpha}x^5 \exp\left(x^2\right) - \alpha t^2 x^4 \exp\left(x^2\right)(5 + 2x^2)$$
$$- 2\alpha \dot{B}(t)t^2 x^3 \exp\left(x^2\right)(10 + 11x^2 + 2x^4).$$

$u(x,t) = \alpha t^2 x^5 \exp(x^2)$ *is the exact solution to the above problem.*

Now, we evaluate $u(x,t)$ along $\bar{P} = 80$ discretized Brownian paths. To approximate $\dot{B}(t)$, we use the discretized scheme described at the beginning of this section. Figure 1 displays the exact and approximate solution with $\alpha = 0.9$ and Figure 2 shows the exact solution and its estimations for different values of α, when $N = 12$. These figures confirm that the resulted numerical solutions have good compatibility with the exact solution. Table 1 displays the l_∞-norm errors and convergence orders for $\alpha = 0.25, 0.75$ and several values of N. Also, Figure 3 show the behaviour of the absolute error of $u(x,t)$ for different values of N, when $\alpha = 0.5$. Table 1 and Figure 3 confirm the accuracy of the obtained numerical approximations.

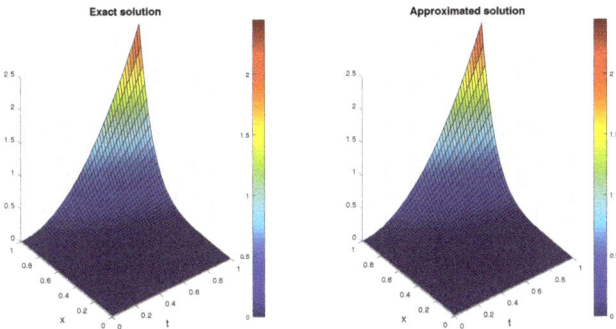

Figure 1. The exact and numerical solution of Example 1 with $\alpha = 0.9$.

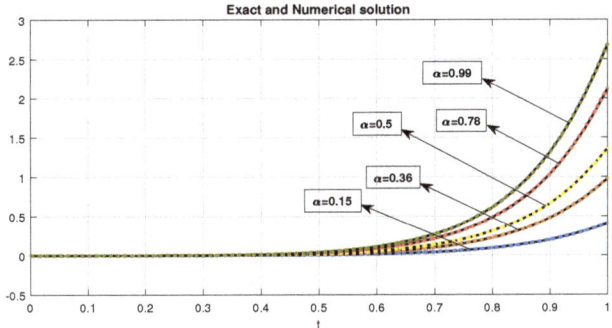

Figure 2. The exact and approximate solution for different values of α in Example 1.

Table 1. Example 1: The l_∞-norm errors and convergence orders.

N	$\alpha = 0.25$		$\alpha = 0.75$	
	$\|\|E_N\|\|_\infty$	CO	$\|\|E_N\|\|_\infty$	CO
6	1.6754×10^{-2}	–	1.1450×10^{-2}	–
9	1.7591×10^{-4}	11.2375	1.8956×10^{-3}	4.4354
12	2.5055×10^{-7}	22.7823	9.5343×10^{-6}	18.3967
15	1.0902×10^{-9}	24.3666	8.4121×10^{-8}	21.1988

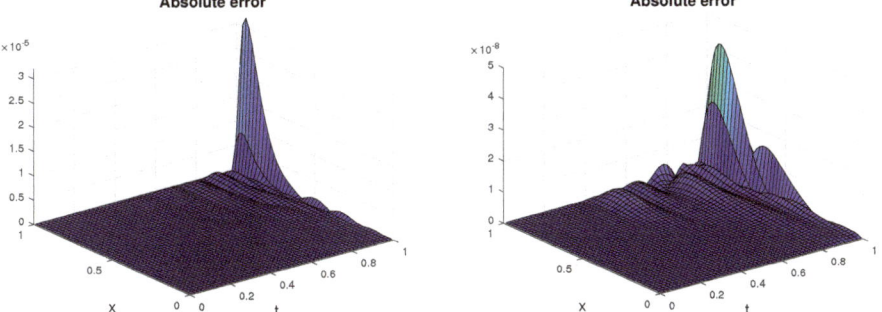

Figure 3. The absolute errors for Example 1 when $N = 12$ (**left**) and $N = 15$ (**right**).

Example 2. *Suppose the time-fractional stochastic equation*

$$D_{0,t}^\alpha u(x,t) = \left(\pi + \dot{B}(t)\right) u_{xx}(x,t) - 2u_x(x,t) + f(x,t),$$

where $\alpha \in (0,1)$, $B(t)$ *is a Brownian motion, and*

$$f(x,t) = \frac{3}{\Gamma(2-\alpha)} t^{1-\alpha} \left(x^2 - \frac{2tx}{\alpha-2} + \frac{2t^2}{(\alpha-2)(\alpha-3)} \right) \sin(\pi x)$$
$$- (\pi + \dot{B}(t))(x+t) \Big[6\sin(\pi x) + 6\pi(x+t)\cos(\pi x)$$
$$- \pi^2(x+t)^2 \sin(\pi x) \Big] + 2(x+t)^2 \Big(\pi(x+t)\cos(\pi x) + 3\sin(\pi x) \Big).$$

With these assumptions, the exact solution is $u(x,t) = (x+t)^3 \sin(\pi x)$.

The numerical solution is evaluated along $\bar{P} = 100$ discretized Brownian paths. Table 2 displays the l_∞-norm errors and order of convergence for several values of α and N. This table shows the high accuracy of the introduced scheme. Also, Figure 4 displays the exact and numerical solution of $u(x,t)$ when $\alpha = 0.5$, $N = 10$ and Figure 5 indicates the absolute error together with the contour plot for $N = 16$. It can be seen that the numerical solution is in well agreement with the exact solution.

Table 2. Example 2: The l_∞-norm errors and convergence order.

N	$\alpha = 0.25$		$\alpha = 0.75$	
	$\|E_N\|_\infty$	CO	$\|E_N\|_\infty$	CO
6	7.6688×10^{-3}	–	6.2094×10^{-3}	–
9	2.8269×10^{-4}	8.1401	1.3207×10^{-4}	9.4964
12	1.7723×10^{-7}	25.6347	1.6552×10^{-8}	23.2270
15	9.9296×10^{-11}	33.5528	7.3368×10^{-11}	34.6026

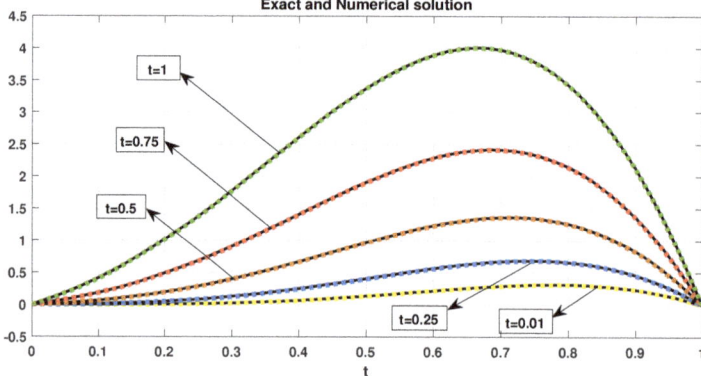

Figure 4. The exact and numerical solution at different levels of t for Example 2.

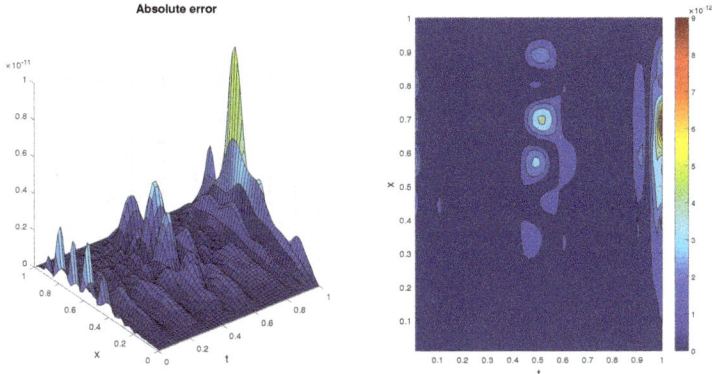

Figure 5. The absolute error (**left**) and contour plot (**right**) for Example 2 with $N = 16$.

Example 3. *Let*

$$D_{0,t}^\alpha u(x,t) = \left(\frac{1}{\pi^2} + \vartheta \dot{B}(t)\right) u_{xx}(x,t),$$

subject to:

$$u(0,t) = u(1,t) = 0,$$
$$u(x,0) = \sin(\pi x),$$

where $\alpha \in (0,1)$ and $B(t)$ is a Brownian motion.

The numerical solutions are evaluated along $\bar{P} = 50$ discretized Brownian paths. Figure 6 shows the numerical solution at $t = 1$ for different values of α, when $\vartheta = 0.5$ and $N = 10$. Figure 7 displays the estimation of $u(x,t)$ when $\vartheta = 0.15, 0.2$, $N = 8$ and $\alpha = 1$. The results are compared with wavelets Galerkin (WG) method [30]. This figure confirms that the present method gives more smooth solution than the numerical scheme in [30]. Also, Figures 8 and 9 indicate the approximate solutions and the contour plots for several values of ϑ when $\alpha = 0.45$. The results confirm that the employed approach is very efficient.

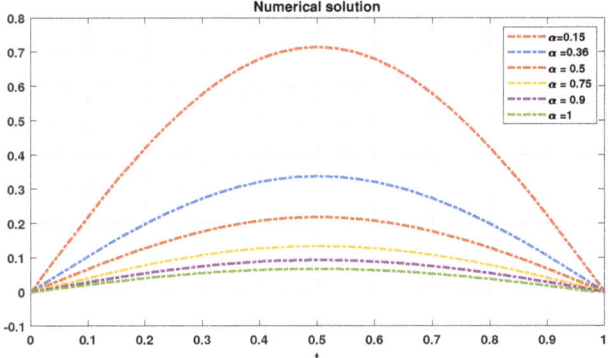

Figure 6. The numerical solution at $t = 1$ for different values of α in Example 3.

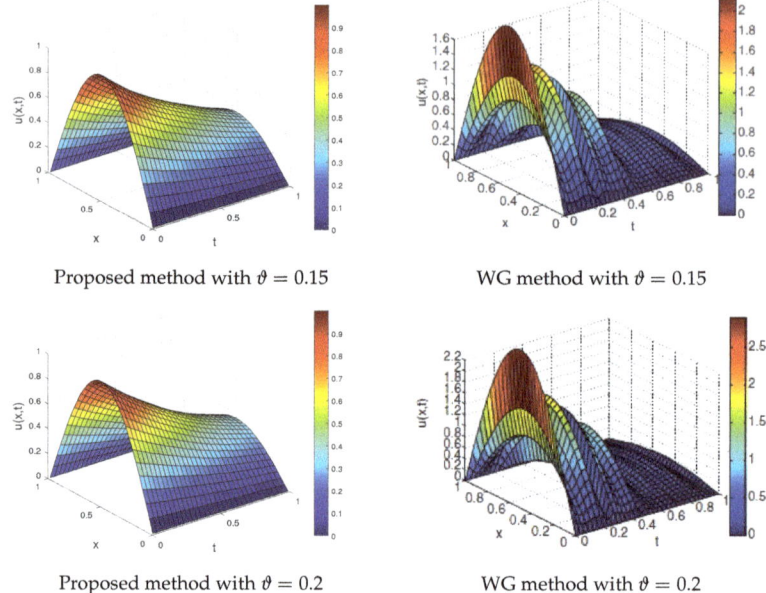

Figure 7. The numerical solution obtained by the proposed method (**left**) and wavelets Galerkin method [30] (**right**) for Example 3 with different values of ϑ when $N = 8$.

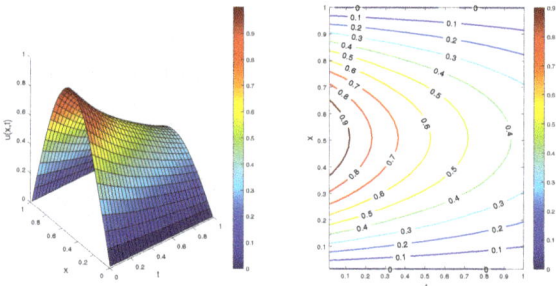

Figure 8. The numerical approximation (**left**) and contour plot (**right**) for Example 3 with $\vartheta = 0.5$.

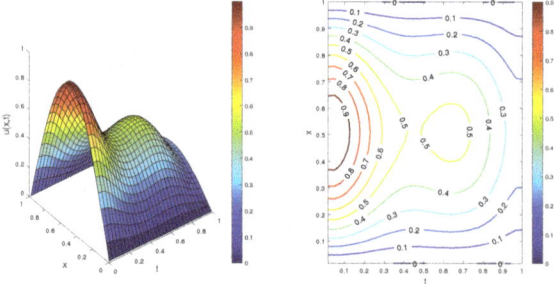

Figure 9. The numerical approximation (**left**) and contour plot (**right**) for Example 3 with $\vartheta = 1.2$.

6. Conclusions

According to numerous applications of FSPDEs, a new numerical scheme was introduced to solve a class of stochastic heat equations of fractional order with additive noise subject to suitable conditions. This numerical method was based on a collocation approach with the SKCPs basis functions. The convergence of the proposed method was proved. Three illustrative examples were investigated to authenticate the efficiency of the discussed approach. The obtained numerical results approved the accuracy of this method.

Author Contributions: All authors discussed the results and contributed to the final manuscript. Also They have read and agreed to the published version of the manuscript.

Funding: This research received no external funding.

Conflicts of Interest: The authors declare no conflict of interest.

References

1. Bellomo, N.; Brzezniak, Z.; Socio, L.M.D. *Nonlinear Stochastic Evolution Problems in Applied Sciences*; Kluwer Academic Publishers, Springer: Dordrecht, The Netherlands, 1992.
2. Brace, A.; Gatarak, D.; Musiela, M. The market model of interest rate dynamics. *Math. Financ.* **1997**, *7*, 127–147. [CrossRef]
3. Singh, S.; Ray, S.S. Numerical solutions of stochastic Fisher equation to study migration and population behavior in biological invasion. *Int. J. Biomath.* **2017**, *10*, 1750103. doi:10.1142/S1793524517501030. [CrossRef]
4. Zmievskaya, G.I.; Bondareva, A.L.; Levchenko, T.V.; Maino, G. Computational stochastic model of ions implantation. *AIP Conf. Proc.* **2015**, *1648*, 230003.
5. Chen, X.; Hu, P.; Shum, S.; Zhang, Y. Dynamic stochastic inventory management with reference price effects. *Oper. Res.* **2016**, *64*, 1529–1536. [CrossRef]
6. Gillard, N.; Belin, E.; Chapeau-Blondeau, F. Stochastic antiresonance in qubit phase estimation with quantum thermal noise. *Phys. Lett. A* **2017**, *381*, 2621–2628. [CrossRef]
7. Kilbas, A.A.; Srivastava, H.M.; Trujillo, J.J. *Theory and Applications of Fractional Differential Equations*; North-Holland Mathematics Studies; Elsevier: Amsterdam, The Netherlands, 2006; Volume 204.
8. Podlubny, I. *Fractional Differential Equations: An Introduction to Fractional Derivatives, Fractional Differential Equations, to Methods of Their Solution and Some of Their Applications*; Academic Press: New York, NY, USA, 1999.
9. Babaei, A.; Banihashemi, S. A stable numerical approach to solve a time-fractional inverse heat conduction problem. *Iran J. Sci. Technol. Trans. A* **2017**, *42*, 2225–2236. [CrossRef]
10. Nemati, S.; Lima, P.M. Numerical solution of nonlinear fractional integro-differential equations with weakly singular kernels via a modification of hat functions. *Appl. Math. Comput.* **2018**, *327*, 79–92. [CrossRef]
11. Babaei, A.; Banihashemi, S. Reconstructing unknown nonlinear boundary conditions in a time-fractional inverse reaction-diffusion-convection problem. *Numer. Methods Part. Differ. Equ.* **2018**, *35*, 976–992. [CrossRef]
12. Tien, D.N. Fractional stochastic differential equations with applications to finance. *J. Math. Anal. Appl.* **2013**, *397*, 334–348. [CrossRef]
13. Sabatelli, L.; Keating, S.; Dudley, J.; Richmond, P. Waiting time distributions in financial markets. *Eur. Phys. J. B Condens. Matter Complex Syst.* **2002**, *27*, 273–275. [CrossRef]
14. Yu, Z.-G.; Anh, V.; Wang, Y.; Mao, D.; Wanliss, J. Modeling and simulation of the horizontal component of the geomagnetic field by fractional stochastic differential equations in conjunction with empirical mode decomposition. *J. Geophys. Res. Space Phys.* **2010**, *115*, doi:10.1029/2009ja015206. [CrossRef]
15. Schumer, R.; Benson, D.A.; Meerschaert, M.M.; Baeumer, B. Multiscaling fractional advection-dispersion equations and their solutions. *Water Resour. Res.* **2003**, *39*, 1022–1032. [CrossRef]
16. Chaves, A.S. A fractional diffusion equation to describe Levy flights. *Phys. Lett. A* **1998**, *239*, 13–16. [CrossRef]
17. Abdel-Rehim, E. From the Ehrenfest model to time-fractional stochastic processes. *J. Comput. Appl. Math.* **2009**, *233*, 197–207. [CrossRef]

18. Chen, Z.Q.; Kim, K.H.; Kim, P. Fractional time stochastic partial differential equations. *Stoch. Process. Appl.* **2015**, *125*, 1470–1499. [CrossRef]
19. Rajivganthi, C.; Muthukumar, P.; Priya, B.G. Successive approximation and optimal controls on fractional neutral stochastic differential equations with poisson jumps. *Optim. Control Appl. Methods* **2015**, *37*, 627–640. [CrossRef]
20. Liu, W.; Tian, K.; Foondun, M. On some properties of a class of fractional stochastic heat equations. *J. Theor. Probab.* **2017**, *30*, 1310–1333. [CrossRef]
21. Ralchenko, K.; Shevchenko, G. Existence and uniqueness of mild solution to fractional stochastic heat equation. *arXiv* **2018**, arXiv:1811.12475.
22. Roozbahani, M.M.; Aminikhah, H.; Tahmasebi, M. Numerical solution of stochastic fractional pdes based on trigonometric wavelets. *UPB Sci. Bull. Ser. A Appl. Math. Phys.* **2018**, *80*, 161–174.
23. Moghaddam, B.P.; Zhang, L.; Lopes, A.M.; Machado, J.A.T.; Mostaghim, Z.S. Sufficient conditions for existence and uniqueness of fractional stochastic delay differential equations. *Int. J. Probab. Stoch. Process.* **2019**. [CrossRef]
24. Mishura, Y.; Ralchenko, K.; Zili, M. On mild and weak solutions for stochastic heat equations with piecewise-constant conductivity. *Stat. Probab. Lett.* **2020**, *159*, 108682. [CrossRef]
25. Gyongy, I.; Martinez, T. On numerical solution of stochastic partial differential equations of elliptic type. *Stochastics* **2006**, *78*, 213–231. [CrossRef]
26. Roth, C. A Combination of Finite Difference and Wong-Zakai Methods for Hyperbolic Stochastic Partial Differential Equations. *Stoch. Anal. Appl.* **2006**, *24*, 221–240. [CrossRef]
27. Walsh, J.B. On numerical solutions of the stochastic wave equation. *Illinois J. Math.* **2006**, *50*, 991–1018. [CrossRef]
28. Du, Q.; Zhang, T. Numerical approximation of some linear stochastic partial differential equations driven by special additive noises. *SIAM J. Numer. Anal.* **2002**, *40*, 1421–1445. [CrossRef]
29. Geissert, M.; Kovacs, M.; Larsson, S. Rate of weak convergence of the finite element method for the stochastic heat equation with additive noise. *BIT* **1995**, *49*, 343–356. [CrossRef]
30. Heydari, M.H.; Hooshmandasl, M.R.; Loghmani, G.B.; Cattani, C. Wavelets Galerkin method for solving stochastic heat equation. *Int. J. Comput. Math.* **2016**, *93*, 1579–1596. [CrossRef]
31. Moghaddam, B.P.; Zhang, L.; Lopes, A.M.; Machado, J.A.T.; Mostaghim, Z.S. Computational scheme for solving nonlinear fractional stochastic differential equations with delay. *Stoch. Anal. Appl.* **2019**, *37*, 893–908. [CrossRef]
32. Mirzaee, F.; Alipour, S. Cubic B-spline approximation for linear stochastic integro-differential equation of fractional order. *J. Comput. Appl. Math.* **2020**, *366*, 112440 [CrossRef]
33. Mirzaee, F.; Hadadiyan, E. Solving system of linear Stratonovich Volterra integral equations via modification of hat functions. *Appl. Math. Comput.* **2017**, *293*, 254–264. [CrossRef]
34. Li, Q.; Kang, T.; Zhang, Q. Mean-square dissipative methods for stochastic age-dependent Capital system with fractional Brownian motion and jumps. *Appl. Math. Comput.* **2018**, *339*, 81–92. [CrossRef]
35. Heydari, M.H.; Mahmoudi, M.R.; Shakiba, A.; Avazzadeh, Z. Chebyshev cardinal wavelets and their application in solving nonlinear stochastic differential equations with fractional Brownian motion. *Commun. Nonlinear Sci. Numer. Simul.* **2018**, *64*, 98–121. [CrossRef]
36. Higham, D.J. An Algorithmic Introduction to Numerical Simulation of Stochastic Differential Equations. *Soc. Ind. Appl. Math.* **2001**, *43*, 525–546. [CrossRef]
37. Liu, L.; Caraballo, T. Well-posedness and dynamics of a fractional stochastic integro-differential equation. *Phys. D Nonlinear Phenom.* **2017**, *355*, 45–57. [CrossRef]
38. Darehmiraki, M. An efficient solution for stochastic fractional partial differential equations with additive noise by a meshless method. *Int. J. Appl. Comput. Math.* **2018**, *4*, 14. [CrossRef]
39. Kovács, M.; Larsson, S.; Saedpanah, F. Mittag–Leffler Euler Integrator for a Stochastic Fractional Order Equation with Additive Noise. *SIAM J. Numer. Anal.* **2020**, *58*, 66–85. [CrossRef]
40. Brzezniak, Z.; Debbi, L. On stochastic Burgers equation driven by a fractional Laplacian and space–time white noise, Stochastic differential equations: Theorem and applications. *Interdiscip. Math. Sci.* **2007**, *2*, 135–167.
41. Shu, J.; Li, P.; Zhang, J.; Liao, O. Random attractors for the stochastic coupled fractional Ginzburg-Landau equation with additive noise. *J. Math. Phys.* **2015**, *56*, 102702. [CrossRef]

42. Meerschaert, M.M.; Tadjeran, C. Finite difference approximations for fractional advection-dispersion flow equations. *J. Comput. Appl. Math.* **2004**, *172*, 65–77. [CrossRef]
43. Masjed-Jamei, M. A basic class of symmetric orthogonal polynomials using the extended Sturm–Liouville theorem for symmetric functions. *J. Math. Anal. Appl.* **2007**, *325*, 753–775. [CrossRef]
44. Babaei, A.; Jafari, H.; Banihashemi, S. Numerical solution of variable order fractional nonlinear quadratic integro-differential equations based on the sixth-kind Chebyshev collocation method. *J. Comput. Appl. Math.* **2020**, *377*, 112908. [CrossRef]
45. Abd-Elhameed, W.M.; Youssri, Y.H. Sixth-kind Chebyshev spectral approach for solving fractional differential equations. *J. Nonlinear Sci. Numer. Simul.* **2019**, *20*, 191–203. [CrossRef]
46. Jafari, H.; Babaei, A.; Banihashemi, S. A novel approach for solving an inverse reaction-diffusion-convection problem. *J. Optim. Theory Appl.* **2019**, *183*, 688–704. [CrossRef]

© 2020 by the authors. Licensee MDPI, Basel, Switzerland. This article is an open access article distributed under the terms and conditions of the Creative Commons Attribution (CC BY) license (http://creativecommons.org/licenses/by/4.0/).

Article

Oscillation Criteria for First Order Differential Equations with Non-Monotone Delays

Emad R. Attia [1,†], Hassan A. El-Morshedy [2,†], and Ioannis P. Stavroulakis [3,4,*]

1 Department of Mathematics, College of Sciences and Humanities, Prince Sattam Bin Abdulaziz University, Alkharj 11942, Saudi Arabia; emadr@du.edu.eg
2 Department of Mathematics, College of Science, University of Bisha, Bisha 61922, P.O. Box 344, Saudi Arabia; elmorshedy@yahoo.com or helmorshedy@ub.edu.sa
3 Department of Mathematics, University of Ioannina, 451 10 Ioannina, Greece
4 Faculty of Mechanics and Mathematics, Al-Farabi Kazakh National University, Almaty 050040, Kazakhstan
* Correspondence: ipstav@uoi.gr
† On leave: Department of Mathematics, Faculty of Science, Damietta University, New Damietta 34517, Egypt.

Received: 18 February 2020; Accepted: 31 March 2020; Published: 2 May 2020

Abstract: New sufficient criteria are obtained for the oscillation of a non-autonomous first order differential equation with non-monotone delays. Both recursive and *lower-upper* limit types criteria are given. The obtained results improve most recent published results. An example is given to illustrate the applicability and strength of our results.

Keywords: Oscillation; Differential equations; Non-monotone delays

1. Introduction

Consider the first order delay differential equation

$$x'(t) + p(t)x(\tau(t)) = 0, \quad t \geq t_0, \tag{1}$$

where $p, \tau \in C([t_0, \infty), [0, \infty))$ and $\tau(t) < t$ for $t \geq t_0$, such that $\lim_{t\to\infty} \tau(t) = \infty$.

A solution of Equation (1) is a function $x(t)$ on $[\bar{t}, \infty)$, where $\bar{t} = \min_{t \geq t_0} \tau(t)$, which is continuously differentiable on $[t_0, \infty)$ and satisfies Equation (1) for all $t \geq t_0$. As customary, a solution of Equation (1) is called oscillatory if it has arbitrarily large zeros. Equation (1) is said to be oscillatory if all its solutions are oscillatory.

The oscillation of Equation (1) has been extensively studied for many decades; see [1–17]. As far as these authors know, the earliest systematic study of the oscillation of Equation (1) was due to Myshkis [14], who proved that Equation (1) is oscillatory when

$$\limsup_{t\to\infty} (t - \tau(t)) < \infty \quad \text{and} \quad \liminf_{t\to\infty} (t - \tau(t)) \liminf_{t\to\infty} p(t) > \frac{1}{e}.$$

In 1972, Ladas et al. [13] proved that Equation (1) is oscillatory if

$$L := \limsup_{t\to\infty} \int_{\tau(t)}^{t} p(s)ds > 1, \tag{2}$$

where the delay $\tau(t)$ is assumed to be a nondecreasing function.

In 1979, Ladas [12] (for Equation (1) with constant delay) and in 1982, Koplatadze and Chanturija [10] established the celebrated oscillation criterion

$$k := \liminf_{t \to \infty} \int_{\tau(t)}^{t} p(s)ds > \frac{1}{e}. \tag{3}$$

The oscillation of Equation (1) has been studied when $0 < k \leq \frac{1}{e}$, $L \leq 1$ and $\tau(t)$ is nondecreasing, see [8,9,15,16] and the references cited therein. In most of these works, the oscillation criteria have been formulated as relations between L and k. For example, Jaroš and Stavroulakis [8], Kon et al. [9], Philos and Sficas [15], and Sficas and Stavroulakis [16] obtained the following criteria, respectively:

$$L > \frac{\ln(\lambda(k)) + 1}{\lambda(k)} - \frac{1 - k - \sqrt{1 - 2k - k^2}}{2},$$

$$L > 2k + \frac{2}{\lambda(k)} - 1,$$

$$L > 1 - \frac{k^2}{2(1-k)} - \frac{k^2}{2}\lambda(k),$$

and

$$L > \frac{\ln \lambda(k) - 1 + \sqrt{5 - 2\lambda(k) + 2k\lambda(k)}}{\lambda(k)}, \tag{4}$$

where $\lambda(k)$ is the smaller real root of the equation $\lambda = e^{\lambda k}$.

The same problem has been considered for Equation (1) with non-monotone delays, see [2,4,11,17–19]. The latter case is much more complicated than the monotone delays case. In fact, according to Braverman and Karpuz ([2], Theorem 1), condition (2) does not need to be sufficient for the oscillation of Equation (1) if $\tau(t)$ is non-monotone. To overcome this difficulty, many authors used a nondecreasing function $\delta(t)$ defined by:

$$\delta(t) = \max_{s \leq t} \tau(s), \quad t \geq t_0; \tag{5}$$

hence, many results were obtained by using techniques similar to those of the monotonic delays case. Most of these results were given by recursive formulas. Next, we give an overview of such results:

In 1994, Koplatadze and Kvinikadze [11] proved the following interesting result which requires the definition of the sequence of functions $\{\psi_i\}_{i=1}^{\infty}$ as follows:

$$\psi_1(t) = 0, \quad \psi_i(t) = e^{\int_{\tau(t)}^{t} p(s)\psi_{i-1}(s)ds}, \quad i = 2, 3, \ldots \tag{6}$$

Theorem 1 ([11]). *Let $j \in \{1, 2, \ldots\}$ exist such that*

$$\limsup_{t \to \infty} \int_{\delta(t)}^{t} p(s) e^{\int_{\delta(s)}^{\delta(t)} p(u)\psi_j(u)du} ds > 1 - c(k), \tag{7}$$

where k, δ, and ψ_j, are defined respectively by (3), (5), and (6) and

$$c(k) = \begin{cases} 0, & \text{if } k > \frac{1}{e}, \\ \frac{1 - k - \sqrt{1 - 2k - k^2}}{2}, & \text{if } 0 \leq k \leq \frac{1}{e}. \end{cases}$$

Then, Equation (1) is oscillatory.

In 2011, Braverman and Karpuz [2] obtained the following sufficient condition for the oscillation of Equation (1),

$$\limsup_{t\to\infty} \int_{\delta(t)}^{t} p(s)\, e^{\int_{\tau(s)}^{\delta(t)} p(u)\,du}\,ds > 1. \tag{8}$$

In 2014, Stavroulakis [17] improved condition (8) to

$$\limsup_{t\to\infty} \int_{\delta(t)}^{t} p(s)\, e^{\int_{\tau(s)}^{\delta(t)} p(u)\,du}\,ds > 1 - \frac{1-k-\sqrt{1-2k-k^2}}{2}. \tag{9}$$

In 2015, Infante et al. [19] proved that Equation (1) is oscillatory if one of the following conditions is satisfied:

$$\limsup_{t\to\infty} \int_{g(t)}^{t} p(s)\, e^{\int_{\tau(s)}^{g(t)} p(u)\, e^{\int_{\tau(u)}^{u} p(v)dv}\,du}\,ds > 1, \tag{10}$$

or

$$\limsup_{\epsilon\to 0^+}\left(\limsup_{t\to\infty} \int_{g(t)}^{t} p(s) e^{(\lambda(k)-\epsilon)\int_{\tau(s)}^{g(t)} p(u)du}\,ds\right) > 1, \tag{11}$$

where $g(t)$ is a nondecreasing function satisfying that $\tau(t) \leq g(t) \leq t$ for all $t \geq t_1$ and some $t_1 \geq t_0$.

In 2016, El-Morshedy and Attia [4] proved that Equation (1) is oscillatory if there exists a positive integer n such that

$$\limsup_{t\to\infty}\left(\int_{g(t)}^{t} q_n(s)ds + c(k^*) e^{\int_{g(t)}^{t}\sum_{i=0}^{n-1} q_i(s)ds}\right) > 1, \tag{12}$$

where $k^* := \liminf_{t\to\infty}\int_{g(t)}^{t} p(s)\,ds$, c, g are defined as before, and $\{q_n(t)\}$ is given by

$$q_0(t) = p(t),\ q_1(t) = q_0(t)\int_{\tau(t)}^{t} q_0(s) e^{\int_{\tau(s)}^{t} q_0(u)du}\,ds,$$

$$q_n(t) = q_{n-1}(t)\int_{g(t)}^{t} q_{n-1}(s) e^{\int_{g(s)}^{t} q_{n-1}(u)du}\,ds,\quad n=2,3,\ldots.$$

Very recently, Bereketoglu et al. [18] proved that Equation (1) oscillates if for some $\ell \in \mathbb{N}$ the following criterion holds

$$\limsup_{t\to\infty}\int_{g(t)}^{t} p(s) e^{\int_{\tau(s)}^{g(t)} P_\ell(u)du}\,ds > 1 - c(k^*), \tag{13}$$

where

$$P_\ell(t) = p(t)\left[1+\int_{g(t)}^{t} p(s) e^{\int_{\tau(s)}^{t} P_{\ell-1}(u)du}\,ds\right],\ P_0(t) = p(t).$$

In this work, we obtain new sufficient criteria of recursive type for the oscillation of Equation (1), when the delay is non-monotone and $k^* \leq \frac{1}{e} < \tilde{L} < 1$, where $\tilde{L} := \limsup_{t\to\infty} \int_{g(t)}^{t} p(s)ds$. In addition, new practical lower limit-upper limit type criteria similar to those in [8,9,15,16] are obtained. These new conditions improve some results in [2,5,8,9,11,13,16–19]. An illustrative example is given to show the strength and applicability of our results.

2. Main Results

Throughout this work, we assume that c, g, k^*, λ, t_1 are defined as above and $g^i(t)$ stands for the *ith* composition of g.

For fixed $n \in \mathbb{N}$, we define $\{R_{m,n}(t)\}$, $\{Q_{m,n}(t)\}$, eventually, as follows:

$$R_{m,n}(t) = 1 + \int_{\tau(t)}^{t} p(s) e^{\int_{\tau(s)}^{t} p(u) Q_{m-1,n}(u) du} ds, \quad m = 1, 2, \ldots,$$

$$Q_{i,j}(t) = e^{\int_{\tau(t)}^{t} p(s) Q_{i,j-1}(s) ds}, \quad i = 1, 2, \ldots, m-1, \; j = 1, 2, \ldots, n$$

where

$$Q_{0,0}(t) = (\lambda(k^*) - \epsilon) \left(1 + (\lambda(k^*) - \epsilon) \int_{\tau(t)}^{g(t)} p(s) ds \right),$$

$$Q_{0,r}(t) = e^{\int_{\tau(t)}^{t} p(s) Q_{0,r-1}(s) ds}, \quad r = 1, 2, \ldots, n$$

$$Q_{i,0}(t) = R_{i,n}, \quad i = 1, 2, \ldots, m-1$$

and $\epsilon \in (0, \lambda(k^*))$.

Lemma 1. *Assume that $x(t)$ is an eventually positive solution of Equation (1). Then,*

$$\frac{x(\tau(t))}{x(t)} \geq R_{m,n}(t),$$

for all sufficiently large t.

Proof. Since $x(t)$ is an eventually positive solution of Equation (1), there exists a sufficiently large $T > t_1$ such that $x(t)$ satisfies eventually

$$x'(t) + p(t) x(g(t)) \leq 0, \quad t > T.$$

Using ([5], Lemma 2.1.2), for sufficiently small $\epsilon > 0$ and sufficiently large t, we have

$$\frac{x(\tau(t))}{x(t)} \geq \frac{x(g(t))}{x(t)} > \lambda(k^*) - \epsilon. \tag{14}$$

On the other hand, dividing both sides of Equation (1) by $x(t)$ and integrating the resulting equation from s to t, $s \leq t$, we obtain

$$x(s) = x(t) e^{\int_s^t p(u) \frac{x(\tau(u))}{x(u)} du}. \tag{15}$$

Therefore,

$$\begin{aligned} x(\tau(t)) &= x(t) e^{\int_{\tau(t)}^{t} p(u) \frac{x(\tau(u))}{x(g(u))} \frac{x(g(u))}{x(u)} du} \\ &\geq x(t) e^{(\lambda(k^*) - \epsilon) \int_{\tau(t)}^{t} p(u) \frac{x(\tau(u))}{x(g(u))} du}. \end{aligned} \tag{16}$$

Integrating Equation (1) from $\tau(\xi)$ to $g(\xi)$,

$$x(g(\xi)) - x(\tau(\xi)) + \int_{\tau(\xi)}^{g(\xi)} p(r) x(\tau(r)) dr = 0.$$

Using (14) as well as the nonincreasing nature of $x(t)$, it follows that

$$x(g(\xi)) - x(\tau(\xi)) + (\lambda(k^*) - \epsilon) x(g(\xi)) \int_{\tau(\xi)}^{g(\xi)} p(r) dr \leq 0.$$

Thus,
$$\frac{x(\tau(\xi))}{x(g(\xi))} \geq 1 + (\lambda(k^*) - \epsilon) \int_{\tau(\xi)}^{g(\xi)} p(r)dr.$$

This together with (16) gives
$$\frac{x(\tau(t))}{x(t)} \geq e^{(\lambda(k^*)-\epsilon)\int_{\tau(t)}^{t} p(u)\left(1+(\lambda(k^*)-\epsilon)\int_{\tau(u)}^{g(u)} p(r)dr\right)du}$$
$$= e^{\int_{\tau(t)}^{t} p(u)Q_{0,0}(u)du} = Q_{0,1}(t). \tag{17}$$

Since (15) implies that $\frac{x(\tau(t))}{x(t)} = e^{\int_{\tau(t)}^{t} p(s)\frac{x(\tau(s))}{x(s)}ds}$, (17) yields
$$\frac{x(\tau(t))}{x(t)} \geq e^{\int_{\tau(t)}^{t} p(s)Q_{0,1}(s)ds} = Q_{0,2}(t).$$

Repeating this process, we arrive at the following inequality
$$\frac{x(\tau(t))}{x(t)} \geq Q_{0,n}(t). \tag{18}$$

On the other hand, by integrating Equation (1) from $\tau(t)$ to t, we have
$$x(t) - x(\tau(t)) + \int_{\tau(t)}^{t} p(s)x(\tau(s))ds = 0. \tag{19}$$

Using (15), we obtain $x(\tau(s)) = x(t)e^{\int_{\tau(s)}^{t} p(u)\frac{x(\tau(u))}{x(u)}du}$. Therefore, (19) implies that
$$\frac{x(\tau(t))}{x(t)} = 1 + \int_{\tau(t)}^{t} p(s)e^{\int_{\tau(s)}^{t} p(u)\frac{x(\tau(u))}{x(u)}du}ds = 0. \tag{20}$$

Now, substituting (18) into (20), we have
$$\frac{x(\tau(t))}{x(t)} \geq 1 + \int_{\tau(t)}^{t} p(s)e^{\int_{\tau(s)}^{t} p(u)Q_{0,n}(u)du}ds = R_{1,n}(t).$$

From the last inequality and (15), we obtain
$$\frac{x(\tau(t))}{x(t)} \geq e^{\int_{\tau(t)}^{t} p(s)R_{1,n}(s)ds} = e^{\int_{\tau(t)}^{t} p(s)Q_{1,0}(s)ds} = Q_{1,1}(t).$$

It follows from this and (15) that
$$\frac{x(\tau(t))}{x(t)} \geq e^{\int_{\tau(t)}^{t} p(s)Q_{1,1}(s)ds} = Q_{1,2}(t).$$

A simple induction implies that
$$\frac{x(\tau(t))}{x(t)} \geq e^{\int_{\tau(t)}^{t} p(s)Q_{1,n-1}(s)ds} = Q_{1,n}(t).$$

Substituting the previous inequality into (20), we get

$$\frac{x(\tau(t))}{x(t)} \geq 1 + \int_{\tau(t)}^{t} p(s) e^{\int_{\tau(s)}^{t} p(u) Q_{1,n}(u) du} ds = R_{2,n}(t).$$

Therefore, by using the same arguments, as before, we obtain

$$\frac{x(\tau(t))}{x(t)} \geq 1 + \int_{\tau(t)}^{t} p(s) e^{\int_{\tau(s)}^{t} p(u) Q_{m-1,n}(u) du} ds = R_{m,n}(t).$$

□

Theorem 2. *Assume that* $k^* \leq \frac{1}{e}$ *and* $m, n \in \mathbb{N}$ *such that*

$$\limsup_{t \to \infty} \int_{g(t)}^{t} p(s) e^{\int_{\tau(s)}^{g(t)} p(u) e^{\int_{\tau(u)}^{u} p(v) R_{m,n}(v) dv} du} ds > 1 - c(k^*). \tag{21}$$

Then, every solution of Equation (1) is oscillatory.

Proof. Assume the contrary, i.e., there exists a non-oscillatory solution $x(t)$. Due to the linearity of Equation (1), one can assume that $x(t)$ is eventually positive. Now, integrating Equation (1) from $g(t)$ to t, we obtain

$$x(t) - x(g(t)) + \int_{g(t)}^{t} p(s) x(\tau(s)) ds = 0. \tag{22}$$

By using (15), it follows that

$$x(\tau(s)) = x(g(t)) e^{\int_{\tau(s)}^{g(t)} p(u) \frac{x(\tau(u))}{x(u)} du}$$
$$= x(g(t)) e^{\int_{\tau(s)}^{g(t)} p(u) e^{\int_{\tau(u)}^{u} p(v) \frac{x(\tau(v))}{x(v)} dv} du}.$$

Therefore, Lemma 1 yields

$$x(\tau(s)) \geq x(g(t)) e^{\int_{\tau(s)}^{g(t)} p(u) e^{\int_{\tau(u)}^{u} p(v) R_{m,n}(v) dv} du}.$$

Substituting into (22), we get

$$x(t) - x(g(t)) + x(g(t)) \int_{g(t)}^{t} p(s) e^{\int_{\tau(s)}^{g(t)} p(u) e^{\int_{\tau(u)}^{u} p(v) R_{m,n}(v) dv} du} ds \leq 0,$$

that is,

$$\int_{g(t)}^{t} p(s) e^{\int_{\tau(s)}^{g(t)} p(u) e^{\int_{\tau(u)}^{u} p(v) R_{m,n}(v) dv} du} ds \leq 1 - \frac{x(t)}{x(g(t))},$$

for sufficiently large t. Therefore,

$$\limsup_{t \to \infty} \int_{g(t)}^{t} p(s) e^{\int_{\tau(s)}^{g(t)} p(u) e^{\int_{\tau(u)}^{u} p(v) R_{m,n}(v) dv} du} ds \leq 1 - \liminf_{t \to \infty} \frac{x(t)}{x(g(t))}.$$

However, $\liminf_{t\to\infty} \frac{x(t)}{x(g(t))} \geq c(k^*)$ (see [5], Lemma 2.1.3). Consequently,

$$\limsup_{t\to\infty} \int_{g(t)}^{t} p(s) e^{\int_{\tau(s)}^{g(t)} p(u) e^{\int_{\tau(u)}^{u} p(v) R_{m,n}(v) dv} du} ds \leq 1 - c(k^*),$$

which contradicts to (21). □

The proofs of the following two results are basically similar to that of Lemma 1 and Theorem 2.

Theorem 3. *Assume that $k^* \leq \frac{1}{e}$ and*

$$\limsup_{t\to\infty} \int_{g(t)}^{t} p(s) e^{(\lambda(k^*)-\epsilon) \int_{\tau(s)}^{g(t)} p(u) du + (\lambda(k^*)-\epsilon)^2 \int_{\tau(s)}^{g(t)} p(u) \int_{\tau(u)}^{g(u)} p(v) dv du} ds > 1 - c(k^*), \tag{23}$$

where $\epsilon \in (0, \lambda(k^*))$. Then, all solutions of Equation (1) oscillate.

Theorem 4. *Assume that $k^* \leq \frac{1}{e}$ and $m, n \in \mathbb{N}$ such that*

$$\limsup_{t\to\infty} \int_{g(t)}^{t} p(s) e^{\int_{\tau(s)}^{g(t)} p(u) R_{m,n}(u) du} ds > 1 - c(k^*). \tag{24}$$

Then, all solutions of Equation (1) oscillate.

Lemma 2. *Let $x(t)$ be an eventually positive solution of Equation (1). Then,*

$$\limsup_{t\to\infty} \left(\int_{g(t)}^{t} p(s) ds + w(g(t)) \int_{g(t)}^{t} p(s) \int_{\tau(s)}^{g(t)} p(u) e^{\int_{\tau(u)}^{g^2(t)} p(v) w(v) dv} du \, ds \right) = 1 - M,$$

where

$$M := \liminf_{t\to\infty} \frac{x(t)}{x(g(t))}, \quad \text{and} \quad w(t) := \frac{x(g(t))}{x(t)}.$$

Proof. The positivity of $x(t)$ implies that $x(t)$ is an eventually non-increasing function. Integrating Equation (1) from $g(t)$ to t, we obtain

$$x(t) - x(g(t)) + \int_{g(t)}^{t} p(s) x(\tau(s)) ds = 0. \tag{25}$$

Since $\tau(s) \leq g(t)$ for $s \leq t$, integrating Equation (1) from $\tau(s)$ to $g(t)$, we have

$$x(\tau(s)) = x(g(t)) + \int_{\tau(s)}^{g(t)} p(u) x(\tau(u)) du.$$

Substituting into (25), we get

$$x(t) - x(g(t)) + x(g(t)) \int_{g(t)}^{t} p(s) ds + \int_{g(t)}^{t} p(s) \int_{\tau(s)}^{g(t)} p(u) x(\tau(u)) du \, ds = 0. \tag{26}$$

It is clear that $\tau(u) \leq g^2(t)$, for $u \leq g(t)$. Therefore, (15) implies that

$$x(\tau(u)) = x(g^2(t))e^{\int_{\tau(u)}^{g^2(t)} p(v)w(v)dv}.$$

From this and (26), it follows that

$$x(t) - x(g(t)) + x(g(t))\int_{g(t)}^{t} p(s)ds + x(g^2(t))\int_{g(t)}^{t} p(s)\int_{\tau(s)}^{g(t)} p(u)e^{\int_{\tau(u)}^{g^2(t)} p(v)w(v)dv} du\, ds = 0.$$

Consequently,

$$\int_{g(t)}^{t} p(s)ds + w(g(t))\int_{g(t)}^{t} p(s)\int_{\tau(s)}^{g(t)} p(u)e^{\int_{\tau(u)}^{g^2(t)} p(v)w(v)dv} du\, ds = 1 - \frac{x(t)}{x(g(t))}.$$

Therefore,

$$\limsup_{t\to\infty}\left(\int_{g(t)}^{t} p(s)ds + w(g(t))\int_{g(t)}^{t} p(s)\int_{\tau(s)}^{g(t)} p(u)e^{\int_{\tau(u)}^{g^2(t)} p(v)w(v)dv} du\, ds\right) = 1 - \liminf_{t\to\infty}\frac{x(t)}{x(g(t))}.$$

□

The proof of the following theorem is a consequence of Lemmas 1, 2, and ([5], Lemmas 2.1.2 and 2.1.3).

Theorem 5. *Assume that* $k^* \leq \frac{1}{e}$ *and* $m, n \in \mathbb{N}$ *such that*

$$\limsup_{t\to\infty}\left(\int_{g(t)}^{t} p(s)ds + (\lambda(k^*) - \epsilon)\int_{g(t)}^{t} p(s)\int_{\tau(s)}^{g(t)} p(u)e^{\int_{\tau(u)}^{g^2(t)} p(v)R_{m,n}(v)dv} du\, ds\right) > 1 - c(k^*),$$

where $\epsilon \in (0, \lambda(k^*))$. *Then, every solution of Equation (1) is oscillatory.*

Theorem 6. *Let* $\tilde{L} := \limsup_{t\to\infty} \int_{g(t)}^{t} p(s)ds < 1$, $0 < k^* \leq \frac{1}{e}$,

$$\int_{g(s)}^{g(t)} p(u)du \geq \int_{s}^{t} p(u)du, \quad \text{for all } s \in [g(t), t], \tag{27}$$

and

$$A := \liminf_{t\to\infty} \int_{\tau(t)}^{g(t)} p(s)ds. \tag{28}$$

If one of the following conditions is satisfied:

(i) $\tilde{L} > \frac{-1 - A\lambda(k^*) + \sqrt{2 + (1 + A\lambda(k^*))^2 + 2k^*\lambda(k^*)}}{\lambda(k^*)}$,

(ii) $\tilde{L} > 1 + k^* + \frac{1}{\lambda(k^*)} + A - \sqrt{\left(1 + k^* + \frac{1}{\lambda(k^*)} + A\right)^2 - 2\left(k^* + \frac{1}{\lambda(k^*)}\right)}$,

then every solution of Equation (1) is oscillatory.

Proof. Assume that Equation (1) has a nonoscillatory solution $x(t)$; as usual, we assume that $x(t)$ is an eventually positive solution. Let

$$I(t) = \int_{g(t)}^{t} p(s)ds + w(g(t)) \int_{g(t)}^{t} p(s) \int_{\tau(s)}^{g(t)} p(u) e^{\int_{\tau(u)}^{g^2(t)} p(v)w(v)dv} du\, ds, \tag{29}$$

where $w(t) = \frac{x(g(t))}{x(t)}$. Therefore,

$$I(t) \geq \int_{g(t)}^{t} p(s)ds + w(g(t)) \left(\int_{g(t)}^{t} p(s) \int_{\tau(s)}^{g(s)} p(u)\, du\, ds + \int_{g(t)}^{t} p(s) \int_{g(s)}^{g(t)} p(u)\, du\, ds \right).$$

In view of [5], Lemma 2.1.2) and (28), for sufficiently small ϵ, we obtain

$$I(t) \geq \int_{g(t)}^{t} p(s)ds + (\lambda(k^*) - \epsilon) \left((A - \epsilon) \int_{g(t)}^{t} p(s)ds + \int_{g(t)}^{t} p(s) \int_{g(s)}^{g(t)} p(u)\, du\, ds \right).$$

By using (27), it follows that

$$I(t) \geq (1 + (\lambda(k^*) - \epsilon)(A - \epsilon)) \int_{g(t)}^{t} p(s)ds + (\lambda(k^*) - \epsilon) \int_{g(t)}^{t} p(s) \int_{s}^{t} p(u)\, du\, ds. \tag{30}$$

However,

$$\int_{g(t)}^{t} p(s) \int_{s}^{t} p(u)\, du\, ds = \frac{1}{2} \left(\int_{g(t)}^{t} p(s)ds \right)^2.$$

Therefore, (30) implies that

$$I(t) \geq (1 + (\lambda(k^*) - \epsilon)(A - \epsilon)) \int_{g(t)}^{t} p(s)ds + \frac{\lambda(k^*) - \epsilon}{2} \left(\int_{g(t)}^{t} p(s)ds \right)^2. \tag{31}$$

On the other hand, from [9], we have

$$\liminf_{t \to \infty} \frac{x(t)}{x(g(t))} \geq 1 - k^* - \frac{1}{\lambda(k^*)}. \tag{32}$$

Therefore, Lemma 2 and (32) imply that $I(t) < k^* + \frac{1}{\lambda(k^*)} + \epsilon$ for sufficiently large t. Thus, (31) yields

$$(1 + (\lambda(k^*) - \epsilon)(A - \epsilon)) \int_{g(t)}^{t} p(s)ds + \frac{\lambda(k^*) - \epsilon}{2} \left(\int_{g(t)}^{t} p(s)ds \right)^2 \leq I(t) < k^* + \frac{1}{\lambda(k^*)} + \epsilon,$$

or equivalently,

$$(\lambda(k^*) - \epsilon)\Lambda^2 + 2(1 + (\lambda(k^*) - \epsilon)(A - \epsilon))\Lambda - 2k^* - \frac{2}{\lambda(k^*)} - 2\epsilon < 0,$$

where

$$\Lambda := \int_{g(t)}^{t} p(s)ds.$$

Then,
$$\Lambda < \frac{-(1+(\lambda(k^*)-\epsilon)(A-\epsilon))+\sqrt{(1+(\lambda(k^*)-\epsilon)(A-\epsilon))^2+2(\lambda(k^*)-\epsilon)\left(k^*+\frac{1}{\lambda(k^*)}+\epsilon\right)}}{\lambda(k^*)-\epsilon}.$$

Thus,
$$\tilde{L} \leq \frac{-(1+(\lambda(k^*)-\epsilon)(A-\epsilon))+\sqrt{(1+(\lambda(k^*)-\epsilon)(A-\epsilon))^2+2(\lambda(k^*)-\epsilon)\left(k^*+\frac{1}{\lambda(k^*)}+\epsilon\right)}}{\lambda(k^*)-\epsilon}.$$

Now, letting $\epsilon \to 0$, we obtain
$$\tilde{L} \leq \frac{-1-A\lambda(k^*)+\sqrt{2+(1+A\lambda(k^*))^2+2k^*\lambda(k^*)}}{\lambda(k^*)}.$$

This completes the proof of case (i).

To prove case (ii), integrating Equation (1) from $g^2(t)$ to $g(t)$, we obtain
$$x(g(t)) - x(g^2(t)) + \int_{g^2(t)}^{g(t)} p(s)x(\tau(s))ds = 0,$$

which, by using the nonincreasing nature of $x(t)$ and the assumption that $\tau(t) \leq g(t)$, implies that
$$x(g(t)) - x(g^2(t)) + x(g^2(t)) \int_{g^2(t)}^{g(t)} p(s)ds \leq 0. \tag{33}$$

In view of (27), we have
$$\int_{g^2(t)}^{g(t)} p(s)ds \geq \int_{g(t)}^{t} p(s)ds.$$

Substituting into (33), it follows that
$$\frac{x(g^2(t))}{x(g(t))} \geq \frac{1}{1 - \int_{g(t)}^{t} p(s)ds}.$$

From this and (29), we obtain
$$I(t) \geq \int_{g(t)}^{t} p(s)ds + \frac{1}{1 - \int_{g(t)}^{t} p(s)ds} \int_{g(t)}^{t} p(s) \int_{\tau(s)}^{g(t)} p(u)du\, ds.$$

Again Lemma 2 and (32) imply for sufficiently small ϵ that
$$\int_{g(t)}^{t} p(s)ds + \frac{1}{1 - \int_{g(t)}^{t} p(s)ds} \int_{g(t)}^{t} p(s) \int_{\tau(s)}^{g(t)} p(u)du\, ds \leq I(t) < k^* + \frac{1}{\lambda(k^*)} + \epsilon. \tag{34}$$

However, as in the proof of case (i), we have

$$\int_{g(t)}^{t} p(s) \int_{\tau(s)}^{g(t)} p(u) du \, ds = \left(\int_{g(t)}^{t} p(s) \int_{\tau(s)}^{g(s)} p(u) \, du \, ds + \int_{g(t)}^{t} p(s) \int_{g(s)}^{g(t)} p(u) \, du \, ds \right)$$

$$\geq \left((A - \epsilon) \int_{g(t)}^{t} p(s) ds + \int_{g(t)}^{t} p(s) \int_{s}^{t} p(u) \, du \, ds \right)$$

$$= (A - \epsilon) \int_{g(t)}^{t} p(s) ds + \frac{1}{2} \left(\int_{g(t)}^{t} p(s) ds \right)^2. \quad (35)$$

Combining the inequalities (34) and (35), we obtain

$$2\Lambda_1 (1 - \Lambda_1) + 2(A - \epsilon) \Lambda_1 + \Lambda_1^2 - 2\alpha(\epsilon)(1 - \Lambda_1) < 0,$$

where

$$\Lambda_1 = \int_{g(t)}^{t} p(s) ds, \quad \alpha(\epsilon) = k^* + \frac{1}{\lambda(k^*)} + \epsilon.$$

Thus,

$$\Lambda_1^2 - 2(1 + \alpha(\epsilon) + A - \epsilon) \Lambda_1 + 2\alpha(\epsilon) > 0,$$

which implies that $\Lambda_1 < 1 + \alpha(\epsilon) + A - \epsilon - \sqrt{(1 + \alpha(\epsilon) + A - \epsilon)^2 - 2\alpha(\epsilon)}$, and hence

$$\tilde{L} = \limsup_{t \to \infty} \int_{g(t)}^{t} p(s) ds \leq 1 + \alpha(\epsilon) + A - \epsilon - \sqrt{(1 + \alpha(\epsilon) + A - \epsilon)^2 - 2\alpha(\epsilon)}.$$

Letting $\epsilon \to 0$, we obtain

$$\tilde{L} \leq 1 + k^* + \frac{1}{\lambda(k^*)} + A - \sqrt{\left(1 + k^* + \frac{1}{\lambda(k^*)} + A\right)^2 - 2\left(k^* + \frac{1}{\lambda(k^*)}\right)}.$$

□

Remark 1.

(i) Condition (27) is satisfied if (see [9,16])

$$p(g(t))g'(t) \geq p(t), \quad \text{eventually for all } t.$$

(ii) It is easy to show that the conclusion of Theorem 6 is valid, if $p(t) > 0$ and condition (27) is replaced by

$$\liminf_{t \to \infty} \frac{p(g(t))g'(t)}{p(t)} = 1.$$

Corollary 1. *Assume that $0 < k \leq \frac{1}{e}$, $L < 1$ and $\tau(t)$ is a nondecreasing continuous function such that*

$$\int_{\tau(s)}^{\tau(t)} p(u) du \geq \int_{s}^{t} p(u) du, \quad \text{for all } s \in [\tau(t), t].$$

If
$$L > \min\left\{\frac{-1+\sqrt{3+2k\lambda(k)}}{\lambda(k)}, 1+k+\frac{1}{\lambda(k)} - \sqrt{1+\left(k+\frac{1}{\lambda(k)}\right)^2}\right\}, \quad (36)$$

then Equation (1) is oscillatory.

Remark 2.

1- Condition (21), with $n = 1$ and $n = 2$, improves conditions (2), (8), (9) and (10), respectively.
2- Condition (23) improves condition (11).
3- Condition (24), with $n = 1$, improves conditions (13) with $\ell = 1$.
4- It is easy to see that

$$\frac{-1+\sqrt{3+2k\lambda(k)}}{\lambda(k)} \le \frac{\ln \lambda(k) - 1 + \sqrt{5 - 2\lambda(k) + 2k\lambda(k)}}{\lambda(k)},$$

for all $\lambda(k) \in [1, e]$. Therefore, condition (36) improves condition (4).

The following example illustrates the applicability and strength of our result.

Example 1. Consider the first order delay differential equation

$$x'(t) + p(t)x(\tau(t)) = 0, \quad t \ge 2, \quad (37)$$

where (See Figure 1)
$$\tau(t) = t - 1 - \alpha \sin^2(\nu\pi(t+\alpha)) + \alpha,$$

and

$$p(t) := \begin{cases} \frac{1}{(1-\alpha)e}, & t \in [2n, 2n+1-\alpha], \\ \frac{1}{\alpha(1-\alpha)}\left(\beta - \frac{1}{e}\right)(t - 2n - 1) + \frac{\beta}{(1-\alpha)}, & t \in [2n+1-\alpha, 2n+1], \\ \frac{\beta}{(1-\alpha)}, & t \in [2n+1, 2n+2-\alpha], \\ \frac{-1}{\alpha(1-\alpha)}\left(\beta - \frac{1}{e}\right)(t - 2n - 2) + \frac{1}{(1-\alpha)e}, & t \in [2n+2-\alpha, 2n+2], \end{cases}$$

where $n \in \mathbb{N}$, $\alpha = 0.0001$, $\beta = 0.505$ and $\nu = 20,000$. Throughout our calculations, we take $g = \delta$. It is clear, from the definition of δ and τ, that

$$t - 1 \le \tau(t) \le \delta(t) \le t - 1 + \alpha.$$

Notice that

$$k^* = k = \liminf_{t\to\infty} \int_{\tau(t)}^{t} p(s)ds = \lim_{n\to\infty} \int_{\tau(2n+1-\alpha)}^{2n+1-\alpha} p(s)ds = \lim_{n\to\infty} \int_{2n}^{2n+1-\alpha} p(s)ds = \frac{1}{e}. \quad (38)$$

Then, $\lambda(k) = e$, and $\frac{1-k-\sqrt{1-2k-k^2}}{2} \approx 0.1365429862$.

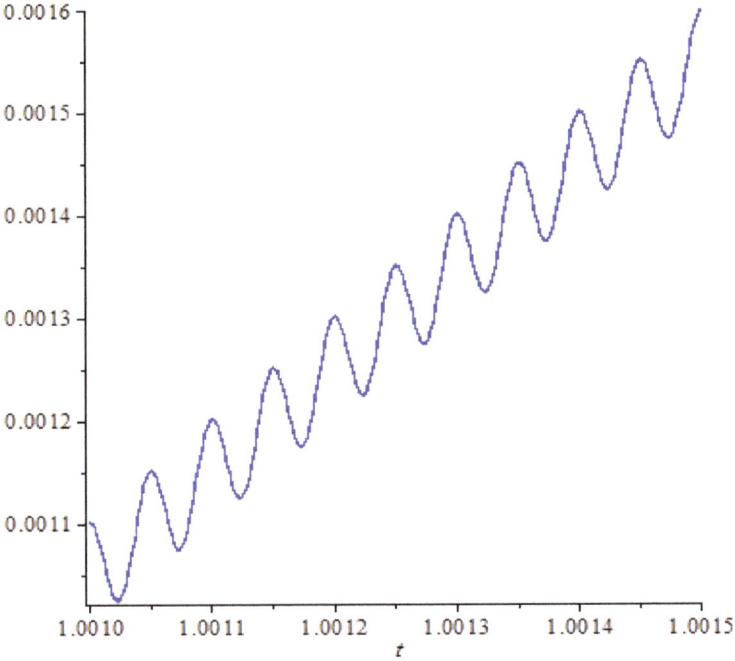

Figure 1. The graph of τ.

Since

$$p(t)R_{1,1}(t) = p(t)\left[1 + \int_{\tau(t)}^{t} p(s) e^{\int_{\tau(s)}^{t} p(u) e^{(\lambda(k)-\epsilon)\int_{\tau(u)}^{u} p(\eta)\left(1+(\lambda(k)-\epsilon)\int_{\tau(\eta)}^{\delta(\eta)} p(r)dr\right)d\eta} du} ds\right],$$

for $\epsilon = 0.0001$, we have

$$p(t)R_{1,1}(t) \geq \frac{1}{(1-\alpha)e}\left[1 + \int_{t-1+\alpha}^{t} \frac{1}{(1-\alpha)e} e^{\int_{s-1+\alpha}^{t} \frac{1}{(1-\alpha)e} e^{(\lambda(k)-\epsilon)\int_{u-1+\alpha}^{u} \frac{1}{(1-\alpha)e} d\eta} du} ds\right] \approx 1.00006322.$$

Now, assume that

$$J(t) = \int_{\delta(t)}^{t} p(s) \exp\left(\int_{\tau(s)}^{\delta(t)} p(u) R_{1,1}(u) du\right) ds.$$

Then,

$$\begin{aligned}
J(2n+2-\alpha) &= \int_{\delta(2n+2-\alpha)}^{2n+2-\alpha} p(s) \exp\left(\int_{\tau(s)}^{\delta(2n+2-\alpha)} p(u) R_{1,1}(u) du\right) ds \\
&\geq \int_{2n+1}^{2n+2-\alpha} p(s) \exp\left(\int_{s-1+\alpha}^{2n+1-\alpha} p(u) R_{1,1}(u) du\right) ds \\
&\geq \int_{2n+1}^{2n+2-\alpha} \frac{\beta}{(1-\alpha)} \exp\left(1.00006322 \int_{s-1+\alpha}^{2n+1-\alpha} du\right) ds \\
&> 0.867626.
\end{aligned}$$

Therefore,

$$\limsup_{t\to\infty} J(t) \geq \lim_{n\to\infty} J(2n+2-\alpha) \geq 0.867626 > 1 - \frac{1-k-\sqrt{1-2k-k^2}}{2} \approx 0.8634570138.$$

Consequently, Theorem (4) with $n = m = 1$ implies that Equation (37) is oscillatory. However, by using (38), condition (3) does not hold.

Let

$$J_1(t) = \int_{\delta(t)}^{t} p(s) \exp\left(\int_{\tau(s)}^{\delta(t)} p(u) \exp\left(\int_{\tau(u)}^{u} p(v) dv\right) du\right) ds.$$

Then,

$$J_1(t) \leq \int_{t-1}^{t} \frac{\beta}{(1-\alpha)} \exp\left(\int_{s-1}^{t-1+\alpha} \frac{\beta}{(1-\alpha)} \exp\left(\int_{u-1}^{u} \frac{\beta}{(1-\alpha)} dv\right) du\right) ds \approx 0.7901391991.$$

Consequently, $\limsup_{t\to\infty} J_1(t) < 0.79014$, which means that conditions (7) with $j = 3$ and (10) fail to apply.

In addition, since

$$\int_{\delta(t)}^{t} p(s) \exp\left(\int_{\tau(s)}^{\delta(t)} p(u) du\right) < \int_{t-1}^{t} \frac{\beta}{(1-\alpha)} \exp\left(\int_{s-1}^{t-1+\alpha} \frac{\beta}{(1-\alpha)} du\right),$$

it follows that

$$\limsup_{t\to\infty} \int_{\delta(t)}^{t} p(s) \exp\left(\int_{\tau(s)}^{\delta(t)} p(u) du\right) < 0.6571023948 < 1 - \frac{1-k-\sqrt{1-2k-k^2}}{2} \approx 0.8634570138.$$

Therefore, none of the conditions (7) with $j = 2$, (8) and (9) are satisfied.

Define

$$J_2(t) = \int_{\delta(t)}^{t} p(s) \int_{\tau(s)}^{s} p(u) \exp\left(\int_{\tau(u)}^{s} p(v) dv\right) du\, ds + c(k) \exp\left(\int_{\delta(t)}^{t} p(s) ds\right).$$

It follows that

$$\begin{aligned}
J_2(t) &\leq \int_{t-1}^{t} \frac{\beta}{(1-\alpha)} \int_{s-1}^{s} \frac{\beta}{(1-\alpha)} \exp\left(\int_{u-1}^{s} \frac{\beta}{(1-\alpha)} dv\right) du\, ds + c(k) \exp\left(\int_{t-1}^{t} \frac{\beta}{(1-\alpha)} ds\right) \\
&< 0.776165,
\end{aligned}$$

so $\limsup_{t\to\infty} J_2(t) \leq 0.776165$. Thus, condition (12) with $n = 1$ fails to apply.

Now, let us define the following functions:

$$J_3(t, \epsilon) = \int_{\delta(t)}^{t} p(s) \exp\left((\lambda(k) - \epsilon) \int_{\tau(s)}^{\delta(t)} p(u) du\right),$$

and

$$J_4(t) = \int_{\delta(t)}^{t} p(s) \exp\left(\int_{\tau(s)}^{\delta(t)} p(u) F_1(u) du\right) ds,$$

where

$$F_1(t) = 1 + \int_{\delta(t)}^{t} p(v) \exp\left(\int_{\tau(v)}^{t} p(u) du\right) dv.$$

Since

$$F_1(t) \leq 1 + \int_{t-1}^{t} \frac{\beta}{1-\alpha} \exp\left(\int_{v-1}^{t} \frac{\beta}{1-\alpha} du\right) dv \approx 2.088615495,$$

and $\lambda(k) - \epsilon < e$, it follows that $J_3(t, \epsilon) < G_e(t)$ and $J_4(t) < G_{2.088615495}(t)$, where $G_\omega(t)$ is defined by

$$G_\omega(t) = \int_{\delta(t)}^{t} p(s) \exp\left(\omega \int_{\tau(s)}^{\delta(t)} p(u) du\right) ds, \quad \text{for } \omega > 0.$$

Next, we estimate the upper limit of $G_\omega(t)$ for $\omega = e$ and $\omega = 2.088615495$.

For $0 \leq \zeta \leq 1 - \alpha$, we have

$$\begin{aligned}
G_\omega(2n + \zeta) &= \int_{\delta(2n+\zeta)}^{2n+\zeta} p(s) \exp\left(\omega \int_{\tau(s)}^{\delta(2n+\zeta)} p(u) du\right) ds \\
&\leq \int_{2n+\zeta-1}^{2n+\zeta} p(s) \exp\left(\omega \int_{s-1}^{2n+\zeta-1+\alpha} p(u) du\right) ds \\
&= \int_{2n+\zeta-1}^{2n-\alpha} p(s) \exp\left(\omega \int_{s-1}^{2n+\zeta-1+\alpha} p(u) du\right) ds \\
&\quad + \int_{2n-\alpha}^{2n} p(s) \exp\left(\omega \int_{s-1}^{2n+\zeta-1+\alpha} p(u) du\right) ds \\
&\quad + \int_{2n}^{2n+\zeta} p(s) \exp\left(\omega \int_{s-1}^{2n+\zeta-1+\alpha} p(u) du\right) ds,
\end{aligned}$$

which implies that

$$\begin{aligned}
G_\omega(2n + \zeta) &\leq \int_{2n+\zeta-1}^{2n-\alpha} \frac{\beta}{(1-\alpha)} \exp\left(\omega \int_{s-1}^{2n-1-\alpha} \frac{1}{(1-\alpha)e} du + \omega \int_{2n-1-\alpha}^{2n+\zeta-1+\alpha} \frac{\beta}{(1-\alpha)} du\right) ds \\
&\quad + \int_{2n-\alpha}^{2n} \frac{\beta}{(1-\alpha)} \exp\left(\omega \int_{s-1}^{2n+\zeta-1+\alpha} \frac{\beta}{(1-\alpha)} du\right) ds \\
&\quad + \int_{2n}^{2n+\zeta} \frac{1}{(1-\alpha)e} \exp\left(\omega \int_{s-1}^{2n+\zeta-1+\alpha} \frac{\beta}{(1-\alpha)} du\right) ds \\
&\approx \frac{1}{\omega} \Big(1.372732323 \, e^{\omega(0.3679804513 + 0.1371342722 \zeta)} - 0.3727323230 \, e^{0.0001010101010 \, \omega (5000 \zeta + 1)} \\
&\quad - e^{0.00005050505050 \, \omega (10000 \zeta + 1)} + 1.980198020 \, e^{0.5050505050 \, \omega \zeta - 1 + 0.00005050505050 \, \omega} \\
&\quad - 1.980198020 \, e^{0.00005050505050 \, \omega - 1} \Big).
\end{aligned}$$

Therefore, $G_{2.088615495}(2n + \zeta) < 0.7725$ and $G_e(2n + \zeta) < 0.9162$ for all $\zeta \in [0, 1 - \alpha]$.

In addition, if $1 - \alpha \leq \zeta \leq 1$, then

$$G_\omega(2n + \zeta) \leq \int_{2n+\zeta-1}^{2n} p(s) \exp\left(\omega \int_{s-1}^{2n+\zeta-1+\alpha} p(u) du\right) ds$$
$$+ \int_{2n}^{2n+\zeta} p(s) \exp\left(\omega \int_{s-1}^{2n+\zeta-1+\alpha} p(u) du\right) ds.$$

Therefore,

$$G_\omega(2n + \zeta) \leq \int_{2n+\zeta-1}^{2n} \frac{\beta}{(1-\alpha)} \exp\left(\omega \int_{s-1}^{2n+\zeta-1+\alpha} \frac{\beta}{(1-\alpha)} du\right) ds$$
$$+ \int_{2n}^{2n+1-\alpha} \frac{1}{(1-\alpha)e} \exp\left(\omega \int_{s-1}^{2n+\zeta-1+\alpha} \frac{\beta}{(1-\alpha)} du\right) ds$$
$$+ \int_{2n+1-\alpha}^{2n+\zeta} \frac{\beta}{(1-\alpha)} \exp\left(\omega \int_{s-1}^{2n+\zeta-1+\alpha} \frac{\beta}{(1-\alpha)} du\right) ds$$
$$\approx \frac{1}{\omega}\left(e^{0.5051010101\omega} - e^{0.00005050505050\omega(10000\zeta+1)} + 1.980198020 e^{-1+0.5050505050\omega\zeta+0.00005050505050\omega}\right.$$
$$\left. - 1.980198020 e^{-1+0.5050505050\omega\zeta-0.5049494949\omega} + e^{0.0001010101010\omega(5000.0\zeta-4999)} - e^{0.00005050505050\omega}\right).$$

Thus, $G_{2.088615495}(2n + \zeta) < 0.6529$ and $G_e(2n + \zeta) < 0.7899$ for all $\zeta \in [1 - \alpha, 1]$.

Using similar arguments, we obtain:

$$G_{2.088615495}(2n + \zeta + 1) < 0.7603, \quad G_e(2n + \zeta) < 0.8737 \text{ for all } \zeta \in [0, 1 - \alpha]$$

and

$$G_{2.088615495}(2n + \zeta + 1) < 0.7603, \quad G_e(2n + \zeta) < 0.8681 \text{ for all } \zeta \in [1 - \alpha, 1].$$

Then,

$$G_{2.088615495}(t) < 0.7725, \quad \text{for all } t \in [2n, 2n + 2], n \in \mathbb{N},$$

and

$$G_e(t) < 0.9162, \quad \text{for all } t \in [2n, 2n + 2], n \in \mathbb{N}.$$

Consequently,

$$\limsup_{\epsilon \to 0^+}\left(\limsup_{t \to \infty} J_3(t, \epsilon)\right) \leq \limsup_{t \to \infty} G_e(t) \leq 0.9162 < 1,$$

and

$$\limsup_{t \to \infty} J_4(t) \leq \limsup_{t \to \infty} G_{2.088615495}(t) \leq 0.7726 < 1 - \frac{1 - k - \sqrt{1 - 2k - k^2}}{2} \approx 0.8634570138.$$

Then, conditions (11) and (13) with $l = 1$ respectively fail to apply.

Author Contributions: All authors contributed equally to the research and to writing the paper. All authors have read and agreed to the published version of the manuscript.

Funding: This research received no external funding.

Acknowledgments: The authors would like to thank the Reviewers for their useful suggestions.

Conflicts of Interest: The authors declare no conflict of interest.

References

1. Agarwal, R.P.; Berezansky, L.; Braverman, E.; Domoshnitsky, A. *Non-Oscillation Theory of Functional Differential Equations with Applications*; Springer: New York, NY, USA; Dordrecht, The Netherlands; Heidelberg, Germany; London, UK, 2012.
2. Braverman, E.; Karpuz, B. On oscillation of differential and difference equations with non-monotone delays. *Appl. Math. Comput.* **2011**, *218*, 3880–3887. [CrossRef]
3. El-Morshedy, H.A. On the distribution of zeros of solutions of first order delay differential equations. *Nonlinear Anal.* **2011**, *74*, 3353–3362. [CrossRef]
4. El-Morshedy, H.A.; Attia, E.R. New oscillation criterion for delay differential equations with non-monotone arguments. *Appl. Math. Lett.* **2016**, *54*, 54–59. [CrossRef]
5. Erbe, L.H.; Kong, Q.; Zhang, B.G. *Oscillation Theory for Functional Differential Equations*; Dekker: New York, NY, USA, 1995.
6. Gopalsamy, K. *Stability and Oscillations in Delay Differential Equations of Population Dynamics*; Kluwer Academic Publishers: Dordrecht, The Netherlands, 1992.
7. Gyori, I.; Ladas, G. *Oscillation Theory of Delay Differential Equations with Applications*; Clarendon Press: Oxford, UK, 1991.
8. Jaroš, J.; Stavroulakis, I.P. Oscillation tests for delay equations. *Rocky Mt. J. Math.* **1999**, *29*, 197–207. [CrossRef]
9. Kon, M.; Sficas, Y.G.; Stavroulakis, I.P. Oscillation criteria for delay equations. *Proc. Am. Math. Soc.* **2000**, *128*, 2989–2997. [CrossRef]
10. Koplatadze, R.G.; Chanturija, T.A. On oscillatory and monotonic solutions of first order differential equations with deviating arguments. *Differential'nye Uravnenija* **1982**, *18*, 1463–1465. (In Russian)
11. Koplatadze, R.G.; Kvinikadze, G. On the oscillation of solutions of first order delay differential inequalities and equations. *Georgian Math. J.* **1994**, *1*, 675–685. [CrossRef]
12. Ladas, G. Sharp conditions for oscillations caused by delays. *Appl. Anal.* **1979**, *9*, 93–98. [CrossRef]
13. Ladas, G.; Lakshmikantham, V.; Papadakis, L.S. *Oscillations of Higher-Order Retarded Differential Equations Generated by the Retarded Arguments, in Delay and Functional Differential Equations and Their Applications*; Academic Press: New York, NY, USA, 1972.
14. Myshkis, A.D. Linear homogeneous differential equations of first order with deviating arguments. *Uspekhi Mat. Nauk* **1950**, *5*, 160–162. (In Russian)
15. Philos, C.G.; Sficas, Y.G. An oscillation criterion for first-order linear delay differential equations. *Canad. Math. Bull.* **1998**, *41*, 207–213. [CrossRef]
16. Sficas, Y.G.; Stavroulakis, I.P. Oscillation criteria for first-order delay equations. *Bull. London Math. Soc.* **2003**, *35*, 239–246. [CrossRef]
17. Stavroulakis, I.P. Oscillation criteria for delay and difference equations with non-monotone arguments. *Appl. Math. Comput.* **2014**, *226*, 661–672. [CrossRef]
18. Bereketoglu, H.; Karakoc, F.; Oztepe, G.S.; Stavroulakis, I.P. Oscillation of first order differential equations with several non-monotone retarded arguments. *Georgian Math. J.* **2019**. [CrossRef]
19. Infante, G.; Koplatadze, R.; Stavroulakis, I.P. Oscillation criteria for differential equations with several retarded arguments. *Funkcial. Ekvac.* **2015**, *58*, 347–364. [CrossRef]

© 2020 by the authors. Licensee MDPI, Basel, Switzerland. This article is an open access article distributed under the terms and conditions of the Creative Commons Attribution (CC BY) license (http://creativecommons.org/licenses/by/4.0/).

Article

Noether Symmetries of a Generalized Coupled Lane-Emden-Klein-Gordon-Fock System with Central Symmetry

B. Muatjetjeja [1,2], S. O. Mbusi [2] and A. R. Adem [3,*]

1 Department of Mathematics, Faculty of Science, University of Botswana, Private Bag 22, Gaborone, Republic of Botswana; ben.muatjetjeja@mopipi.ub.bw
2 Department of Mathematical Sciences, North-West University, Mafikeng Campus, Private Bag X 2046, Mmabatho 2735, South Africa; siveoscar@gmail.com
3 Department of Mathematical Sciences, University of South Africa, Unisa 0003, South Africa
* Correspondence: ademar@unisa.ac.za

Received: 20 February 2020; Accepted: 20 March 2020; Published: 5 April 2020

Abstract: In this paper we carry out a complete Noether symmetry analysis of a generalized coupled Lane-Emden-Klein-Gordon-Fock system with central symmetry. It is shown that several cases transpire for which the Noether symmetries exist. Moreover, we derive conservation laws connected with the admitted Noether symmetries. Furthermore, we fleetingly discuss the physical interpretation of the these conserved vectors.

Keywords: Lane-Emden-Klein-Gordon-Fock system with central symmetry; Noether symmetries; conservation laws

MSC: 35J47; 35G50; 35J61

1. Introduction

In 2017 [1], the authors studied both Lie and Noether symmetries of a Lane-Emden-Klein-Fock system with central symmetry with power functions namely,

$$u_{tt} - u_{rr} - \frac{n}{r}u_r + \frac{\gamma v^q}{r^n} = 0,$$
$$v_{tt} - v_{rr} - \frac{n}{r}v_r + \frac{\alpha u^p}{r^n} = 0, \qquad (1)$$

where p, n, γ, α, q are non-zero constants. In fact, when $n = 2, \gamma = \alpha = 1$, system (1) becomes

$$u_{tt} - u_{rr} - \frac{2}{r}u_r + \frac{v^q}{r^2} = 0,$$
$$v_{tt} - v_{rr} - \frac{2}{r}v_r + \frac{u^p}{r^2} = 0. \qquad (2)$$

System (2) has been studied in [2] for both Lie and Noether symmetries together with the associated conservation laws.

Systems of this type occur in several physical phenomena, see, for example, Refs. [1–4] and references therein. These type of system can also be viewed as a natural extension of the famous two-component generalization of the nonlinear wave equation, viz,

$$u_{tt} - u_{rr} - \frac{m}{r}u_r - u^p = 0, \qquad (3)$$

with the real-valued function $u = u(t,r)$, and p representing the interaction power while the independent variables (t,r) symbolize time and radial coordinates respectively in $m \neq 0$ dimensions [4].

In 2019 [5], the authors studied the generalization of system (1) where the power functions v^q and u^p are replaced with arbitrary elements namely, $h(v)$ and $g(u)$ respectively. Thus system (1) becomes

$$u_{tt} - u_{rr} - \frac{n}{r}u_r + \frac{h(v)}{r^n} = 0,$$

$$v_{tt} - v_{rr} - \frac{n}{r}v_r + \frac{g(u)}{r^n} = 0. \tag{4}$$

It is worth mentioning that, if the parameter $n = 0$ in system (1), then system (1) reduces to the Lane-Emden system

$$u_{xx} + u_{yy} + v^p = 0,$$

$$v_{xx} + v_{yy} + u^q = 0, \tag{5}$$

under the complex transformation $(x, y, u, v) \to (t, ir, u, v)$, where p and q are non-zero constants. This system has been extensively studied for its Noether and Lie symmetries [6]. Furthermore, if the parameter $n = 0$, in system (4), then system (4) transforms to a generalized Lane-Emden system

$$u_{xx} + u_{yy} + h(v) = 0,$$

$$v_{xx} + v_{yy} + g(u) = 0, \tag{6}$$

under the aforementioned complex transformation. In [7], authors applied the classical symmetry method to investigate the symmetries of system (6).

In [5], the authors applied the method of modern group analysis to study a generalized coupled Lane-Emden-Klein-Gordon-Fock system with central symmetry (4). Motivated by the recent results in [5], we study the aforemention system (4). To the authors' knowledge, the method of Noether symmetry analysis has not been used in the study of a generalized Lane-Emden-Klein-Fock system with central symmetry (4). Thus in this paper, we aim to compensate for this absence by carrying out a complete Noether symmetry classification of system (4) and derive the connected conservation laws of system (4). Since system (4), has a Lagrangian structure, thus the knowledge of Noether theorem [8] gives us an elegant way to construct conservation of system (4).

The structure of this paper is as follows. Firstly, we seek to establish the admitted Noether symmetries of a generalized coupled Lane-Emden-Klein-Gordon-Fock system with central symmetry (4) associated with the standard Lagrangian. Next, in Section 2, conservation laws connected with the admitted Noether symmetries are derived. Concluding remarks are summarised in Section 3.

2. Complete Noether Symmetries Analysis

Several authors have done much work on Noether classification for a system of PDEs. See for example [6,7,9]. Here we perform a complete Noether symmetry analysis of system (4) with respect to the standard Lagrangian. System (4) has a Lagrangian structure. This prompts the following Lemma.

Lemma 1. *The generalized coupled Lane-Emden-Klein-Gordon-Fock system with central symmetry (4) establishes the Euler-Lagrange equations with the functional*

$$J(u,v) = \int_0^\infty \int_0^\infty L(t, r, u, v, u_t, v_t, u_r, v_r) dt dr,$$

where

$$\mathcal{L} = \frac{1}{n}\left(r^n u_t v_t - r^n u_r v_r - \int h(v) dv - \int g(u) du\right). \tag{7}$$

is the connected function of Lagrange.

Proof. The insertion of $\mathcal{L}$ in the Euler-Lagrange equations [6,9] gives

$$\begin{aligned}
\frac{\delta \mathcal{L}}{\delta u} &= \frac{\partial \mathcal{L}}{\partial u} - D_t\left(\frac{\partial \mathcal{L}}{\partial u_t}\right) - D_r\left(\frac{\partial \mathcal{L}}{\partial u_r}\right), \\
&= -\frac{g(u)}{n} - \frac{1}{n}D_t(r^n v_t) - \frac{1}{n}D_r(-r^n v_r), \\
&= -\frac{g(u)}{n} - \frac{r^n}{n}v_{tt} - \frac{1}{n}(-nr^{n-1}v_r - r^n v_{rr}), \\
&= v_{tt} - v_{rr} - \frac{n}{r}v_r + \frac{g(u)}{r^n} = 0, \\
\frac{\delta \mathcal{L}}{\delta v} &= \frac{\partial \mathcal{L}}{\partial v} - D_t\left(\frac{\partial \mathcal{L}}{\partial v_t}\right) - D_r\left(\frac{\partial \mathcal{L}}{\partial v_r}\right), \\
&= -\frac{h(v)}{n} - \frac{1}{n}D_t(r^n u_t) - \frac{1}{n}D_r(-r^n u_r), \\
&= -\frac{h(v)}{n} - \frac{r^n}{n}u_{tt} - \frac{1}{n}(-nr^{n-1}u_r - r^n u_{rr}), \\
&= u_{tt} - u_{rr} - \frac{n}{r}u_r + \frac{h(v)}{r^n} = 0,
\end{aligned}$$

Hence this complete the proof. □

Let $x = (x^1, \cdots, x^n)$ be n independent variables and $u = (u^1, \cdots, u^m)$ m dependent variables. An operator (the sum over repeated indices is presupposed)

$$X = \xi^i(x,u)\frac{\partial}{\partial x^i} + \eta^\alpha(x,u)\frac{\partial}{\partial u^\alpha} \tag{8}$$

is called *Noether point symmetry generator* of the coupled system (4) connected to the Lagrangian $\mathcal{L}$ in (7) if the Killing-type equation,

$$X^{(1)}\mathcal{L} + D_i(\xi^i)\mathcal{L} = D_i A^i, \tag{9}$$

holds for some point-dependent potential terms $A = (A^i)$ where $A^i = A^i(t,r,u,v)$, $i = 1,2$. We now revisit the celebrated Noether Theorem [6,8], that is, corresponding to each Noether symmetry, there exist a vector $T = (T^i)$ with components

$$T^i = \xi^i \mathcal{L} + \frac{\partial \mathcal{L}}{\delta u_i^j}(\eta_j - u_s^j \xi^s) - A^i, \tag{10}$$

which is a conserved vector of system (4). The solution of (9) leads to overdetermining systems of PDEs. Solving the resulting systems of PDEs prompts the following results.

$$\begin{aligned}
\tau &= a(t,r), \\
\xi &= b(t,r), \\
\eta^1 &= -d_v(t,r,v)u - \frac{n}{r}bu + k(t,r), \\
\eta^2 &= d(t,r,v), \\
A^1 &= \frac{r^n}{n}d_t u + \frac{r^n}{n}k_t v + s(t,r), \\
A^2 &= -\frac{r^n}{n}d_r u - \frac{r^n}{n}k_r v + w(t,r),
\end{aligned}$$

$$(d_v u - k)g(u) + \frac{n}{r}bug(u) - df(v) - (b_r + a_t)\left[\int h(v)dv + \int g(u)du\right]$$
$$= r^n(d_{tt} - d_{rr})u + r^n(k_{tt} - k_{rr})v - nr^{n-1}(d_r u + k_r v) + n(s_t + w_r). \quad (11)$$

A complete analysis of Equation (11) yields the following results.

Theorem 1. *Suppose $n \neq 0$, $h(v)$ and $g(u)$ are arbitrary functions, then the Noether generator of a generalized coupled Lane-Emden-Klein-Gordon-Fock system with central symmetry (4) and the associated conservation laws are given by (12)*

$$\begin{cases} X_1 = \frac{\partial}{\partial t}, \\ A^i = 0, \\ T_1^1 = -\frac{r^n}{n}u_t v_t - \frac{r^n}{n}u_r v_r - \frac{1}{n}\int h(v)dv - \frac{1}{n}\int g(u)du, \\ T_1^2 = \frac{r^n}{n}u_t v_r + \frac{r^n}{n}u_r v_t. \end{cases} \quad (12)$$

Theorem 2. *Let the elements $h(v) = \alpha v + \beta$, $g(u) = \gamma u + \lambda$, with $\alpha, \gamma, \beta, \lambda$ are constants, $\alpha, \gamma \neq 0$ and n arbitrary. Then the Noether symmetries of system (4) and the connected conserved vectors are (12) and*

$$\begin{cases} X_2 = k(t,r)\frac{\partial}{\partial u} + f(t,r)\frac{\partial}{\partial v}, \quad A^1 = \frac{r^n}{n}uf_t + \frac{r^n}{n}uk_t, \quad A^2 = -\frac{r^n}{n}uf_r - \frac{r^n}{n}uk_r, \\ \text{with} \quad k_{tt} - k_{rr} - \frac{n}{r}k_r + \frac{\alpha}{r^n}f = 0, \quad f_{tt} - f_{rr} - \frac{n}{r}f_r + \frac{\gamma}{r^n}k = 0, \\ T_2^1 = \frac{r^n}{n}fu_t + \frac{r^n}{n}kv_t - \frac{r^n}{n}uf_t - \frac{r^n}{n}vk_t, \\ T_2^2 = \frac{r^n}{n}uf_r + \frac{r^n}{n}vk_r - \frac{r^n}{n}fu_r - \frac{r^n}{n}kv_r. \end{cases} \quad (13)$$

Theorem 3. *Suppose that $h(v) = \gamma v^q$, $g(u) = \alpha u^p$, $\alpha, \gamma \neq 0$. Then the Noether operators of system (4) and the associated conservation laws are as follows;*

(i) *if $n = \dfrac{2(q+p+2)}{(p+1)(q+1)}$, $p, q \neq 0, \pm 1$, then we have (12) and*

$$\begin{cases} X_2 = t\frac{\partial}{\partial t} + r\frac{\partial}{\partial r} - \frac{2}{p+1}u\frac{\partial}{\partial u} - \frac{2}{q+1}v\frac{\partial}{\partial v}, \\ A^i = 0, \\ T_2^1 = -\frac{tr^n}{n}(u_t v_t + u_r v_r) - \frac{1}{n(p+1)}(\alpha t u^{p+1} + 2r^n u v_t) - \frac{1}{n(q+1)}(\gamma t v^{q+1} + 2r^n u_t v) - \\ \frac{r^{n+1}}{n}(v_t u_r + u_t v_r), \\ T_2^2 = -\frac{r^{n+1}}{n}(u_t v_t + u_r v_r) - \frac{1}{n(p+1)}(\alpha r u^{p+1} - 2r^n u v_r) - \frac{1}{n(q+1)}(\gamma r v^{q+1} - 2r^n u_r v) + \\ \frac{tr^n}{n}(v_t u_r + u_t v_r). \end{cases}$$

(ii) *if $p = q = -1$, $\gamma = \alpha$, n arbitrary. Here we get the generic case (12) and*

$$\begin{cases} X_3 = u\frac{\partial}{\partial u} - v\frac{\partial}{\partial v}, \\ A^i = 0, \\ T_3^1 = \frac{r^n}{n}(uv_t - u_t v), \\ T_3^2 = \frac{r^n}{n}(u_r v - uv_r). \end{cases}$$

It should be noted that in any other case one recovers (12). It should also be observed that when $p = q = 1$, this falls into Theorem 2.

Theorem 4. *Let the elements $h(v) = \alpha v^p$, $g(u) = \gamma e^{-mu}$, $\alpha, \gamma, m \neq 0$, $p \neq -1$. Then the Noether generators of system (4) and the corresponding conservation laws are;*

(i) if $n = \dfrac{2}{p+1}$, $\gamma = \alpha$, n, m arbitrary. Here the generic case (12) extends by one Noether generator with the associated conservation laws;

$$\begin{cases} X_2 = tm(p+1)\dfrac{\partial}{\partial t} + rm(p+1)\dfrac{\partial}{\partial r} + 2(p+1)\dfrac{\partial}{\partial u} - 2mv\dfrac{\partial}{\partial v}, \\ A^i = 0, \\ T_2^1 = -\dfrac{r^n}{n}mtu_r v_r - \dfrac{r^n}{n}mtu_t v_t - \dfrac{m\alpha}{n(p+1)}tv^{p+1} + \dfrac{\gamma t}{n}e^{-mu} + \dfrac{2r^n}{n}v_t - \dfrac{mr^{n+1}}{n}u_r v_t - mr^n v u_t - \dfrac{mr^{n+1}}{n}v_r u_t, \\ T_2^2 = \dfrac{r^{n+1}}{n}mu_t v_t + \dfrac{r^{n+1}}{n}mu_r v_r - \dfrac{m\alpha}{n(p+1)}rv^{p+1} + \dfrac{\gamma}{n}re^{-mu} - \dfrac{2r^n}{n}v_r + \dfrac{r^n}{n}mtu_t v_r + r^n m v u_r + \dfrac{r^n}{n}mtv_t u_r. \end{cases}$$

It should be noted that in any other case one recovers (12). This analysis will also be encountered in Theorem 5.

Theorem 5. *Suppose that $h(v) = \alpha e^{\lambda v}$, $g(u) = \gamma u^q$, $q, \gamma, \alpha \neq 0$. Then the Noether operators of system (4) and the associated conserved vectors are;*

(i) if $n = \dfrac{2}{q+1}$, $\gamma = \alpha$, n, λ arbitrary. In this case, the generic case (12) enlarges by one operator with the following conserved vectors;

$$\begin{cases} X_2 = t\lambda(q+1)\dfrac{\partial}{\partial t} + r\lambda(q+1)\dfrac{\partial}{\partial r} + 2(q+1)\dfrac{\partial}{\partial u} - 2\lambda v\dfrac{\partial}{\partial v}, \\ A^i = 0, \\ T_2^1 = -\dfrac{r^n}{n}\lambda t u_r v_r - \dfrac{r^n}{n}\lambda t u_t v_t - \dfrac{\gamma\lambda}{n(q+1)}tu^{p+1} + \dfrac{\alpha t}{n}e^{-\lambda v} + \dfrac{2r^n}{n}u_t - \dfrac{\lambda r^{n+1}}{n}u_r v_t - \lambda r^n u v_t - \dfrac{\lambda r^{n+1}}{n}v_r u_t, \\ T_2^2 = \dfrac{r^{n+1}}{n}\lambda u_t v_t + \dfrac{r^{n+1}}{n}\lambda u_r v_r - \dfrac{\lambda\gamma}{n(q+1)}ru^{q+1} + \dfrac{\alpha}{n}re^{-\lambda v} - \dfrac{2r^n}{n}u_r + \dfrac{r^n}{n}\lambda t v_t u_r + r^n \lambda u v_r + \dfrac{r^n}{n}\lambda t u_t v_r. \end{cases}$$

It should be observed that in any other case one recovers (12).

The aforementioned theorems can be proved by inserting the values of X_i, n, $h(v)$ and $g(u)$ into Equation (11) and these will satisfy Equation (11). Moreover, substituting these values into Equation (10) one obtains the associated T^i. These T^i then satisfy the divergence condition.

Remark 1. *It is worth mentioning that for any case that do not fall in Theorems 2–5, the Noether algebra is one-dimentional and is generated by X_1. It should be noted that Theorem 2 cannot be directly obtained as a consequence of the results of [1], since the functions $h(v)$ and $g(u)$ are not linear, but affine functions, hence these give some new results. In addition, Theorems 4 and 5 exploit new forms of $h(v)$ and $g(u)$ which also lead to some new results. The cases when $h(v)$ and $g(u)$ are constants are discarded.*

3. Concluding Remarks

A complete Noether symmetry classification of the generalized coupled Lane-Emden-Klein-Gordon-Fock system with central symmetry (4) was carried out. Several functional forms of the elements $h(v)$ and $g(u)$ which resulted in Noether point symmetries were derived. Thereafter, conservation laws connected to the Noether point symmetries were obtained. Conservation laws are of undisputed significance. From the mathematical point of view, when analyzed, they can be employed to detect integrability. Although conservation laws are useful in the analysis of solutions of differential equations, we will exclude this analysis for our future work. The results of the problem under study were motivated by the recent work in [1]. However, the results derived therein were not complete since the function $h(v)$ and $g(u)$ were only considered to be power functions. However, in the present work, the function $h(v)$ and $g(u)$ were consider to be arbitrary, and this resulted in some new and more general results. The authors thank the anonymous referees whose comments helped to improve the paper.

Author Contributions: Conceptualization, B.M., Conceptualization, S.O.M., Conceptualization, A.R.A. All authors have read and agreed to the published version of the manuscript.

Funding: This research received no external funding.

Acknowledgments: The authors thank the anonymous referees whose comments helped to improve the paper.

Conflicts of Interest: The authors declare no conflict of interest.

References

1. Muatjetjeja, B. Coupled Lane-Emden-Klein-Gordon-Fock system with central symmetry: Symmetries and Conservation laws. *J. Differ. Equ.* **2017**, *263*, 8322–8328. [CrossRef]
2. Freire, I.L.; Muatjetjeja, B. Symmetry analysis of a Lane-Emden-Klein-Gordon-Fock system with central symmetry. *Discret. Contin. Dyn. Syst. Ser. A* **2018**, *11*, 667–673. [CrossRef]
3. Muatjetjeja, B.; Mogorosi, T.E. Variational principle and conservation laws of a generalized hyperbolic Lane-Emden system. *J. Comput. Nonlinear Dyn.* **2018**, *13*, 121002. [CrossRef]
4. Mogorosi, T.E.; Freire, I.L.; Muatjetjeja, B.; Khalique, C.M. Group anylsis of a hyperbolic Lane-Emden system. *Appl. Math. Comput.* **2017**, *292*, 156–164.
5. Mbusi, S.O.; Muatjetjeja, B.; Adem, A.R. Lie group classification of a generalized coupled Lane-Emden-Klein-Gordon-Fock system with central symmetry. *Nonlinear Dyn. Syst. Theory* **2019**, *19*, 186–192.
6. Bozhkov, Y.; Freire, I.L. Symmetry analysis of the bidimensional Lane-Emden system. *J. Math. Anal. Appl.* **2012**, *388*, 1279–1284. [CrossRef]
7. Muatjetjeja, B.; Khalique, C.M. Conservation laws for a generalized coupled bidimensional Lane-Emden system. *Commun. Nonlinear Sci. Numer. Simulat.* **2013**, *18*, 851–857. [CrossRef]
8. Noether, E. Invariante Variationsprobleme. *Nachrichten von der Gesellschaft der Wissenschaften zu Göttingen, Mathematisch-Physikalische Klasse* **1918**, *2*, 235–257.
9. Bozhkov, Y.; Martins, A.C.G. Lie point symmetries of the Lane-Emden system. *J. Math. Anal. Appl.* **2004**, *294*, 334–344. [CrossRef]

© 2020 by the authors. Licensee MDPI, Basel, Switzerland. This article is an open access article distributed under the terms and conditions of the Creative Commons Attribution (CC BY) license (http://creativecommons.org/licenses/by/4.0/).

Article

Finite Difference Approximation Method for a Space Fractional Convection–Diffusion Equation with Variable Coefficients

Eyaya Fekadie Anley [1,2] and Zhoushun Zheng [1],*

[1] School of Mathematics and Statistics, Central South University, Changsha 410083, China; eyayafek@csu.edu.cn

[2] College of Natural and Computational Science, Department of Mathematics, Arba-Minch University, Arba-Minch 21, Ethiopia

* Correspondence: 2009zhengzhoushun@163.com

Received: 22 February 2020; Accepted: 1 March 2020; Published: date

Abstract: Space non-integer order convection–diffusion descriptions are generalized form of integer order convection–diffusion problems expressing super diffusive and convective transport processes. In this article, we propose finite difference approximation for space fractional convection–diffusion model having space variable coefficients on the given bounded domain over time and space. It is shown that the Crank–Nicolson difference scheme based on the right shifted Grünwald–Letnikov difference formula is unconditionally stable and it is also of second order consistency both in temporal and spatial terms with extrapolation to the limit approach. Numerical experiments are tested to verify the efficiency of our theoretical analysis and confirm order of convergence.

Keywords: Crank–Nicolson scheme; Shifted Grünwald–Letnikov approximation; space fractional convection-diffusion model; variable coefficients; stability analysis

MSC: 26A33; 35R11; 65L20

---

1. Introduction

Fractional differential equations (FDE) have attracted the attention of many researchers and scientists due to their importance in different fields of study such as viscoelasticity, fluid mechanics, physics, biology, engineering, and flows in porous media (see [1–6] and the references cited therein). As different experiments and implementations have shown, non-integer space derivatives have been used to develop anomalous diffusion to which a particle spreads at a rate inconsistent with the integer Brownian motion problem in the direction of both time and space. When non-integer order is replaced by the second order derivative in a diffusion equation, it acts to enhance the process which we call super-diffusion [7–12]. Laboratory experiments and field-scale tracer dispersion breakthrough curves (BTCs) are suitable for exhibiting early time arrivals that are not captured by the integer order derivatives and these non-Fickian phenomena can be controlled by non-classical order convection–diffusion and dispersion equations (FCDE) as it was explained in [13]. To increase the number of applications, there should be significant interest in constructing numerical schemes to solve a well known space fractional convection–diffusion model that has space variable coefficients. In most cases, non-integer order differential problems have no exact solution, so various iterative and numerical approximations [3,9,14] must be pointed out in advance. In general, these kinds of approaches have become important in finding the approximate solutions of fractional differential equations, so extensive numerical methods have been developed for space fractional convection–diffusion equations such as the spectral method [15], finite volume method [16,17], finite difference method [2,9,14,18–26], finite element method [27–30] and collocation method [31,32].

When the discretization of domain over the region (which belongs to the geometry) is not complex, finite difference approximations are easier and faster than other methods (see [16,33] for further details) to get numerical solutions. In [34], the author used an unconditional stable difference method for time–space fractional convection–diffusion problems with space variable coefficients with first order convergence both in time and space. The Crank–Nicolson finite difference method for one-sided space fractional diffusion equations using an extrapolation method to get second order convergence was studied in [23]. In [9], the explicit and implicit finite difference methods are discussed for a one-sided space fractional convection–diffusion equation with first order convergence in both time and space. A first-order implicit finite difference discretization method for a two-sided space fractional diffusion equation (SFDE) is also applied in [10]. Recently, an unconditionally stable second order accurate difference method for a two-sided time–space fractional convection–diffusion equation was constructed in [35] using the weighted and Shifted Grünwald–Letnikov difference approximation. It is not suitable to apply the weighted combined with shifted Grünwald–Letnikov difference approximation for one-sided Riemann–Liouville fractional derivative to have second order accurate in space. To deal with such issues, it is important to develop a numerical scheme that leads to evaluate a one-sided space fractional convection–diffusion problem. Thus, the main focus of our study is to have temporal and spatial second order convergence estimates for one-sided space fractional convection–diffusion equations based on a stable finite difference method and using spatial extrapolation to the limit approach. The scheme has been treated using the Crank–Nicholson method with the novel Shifted Grünwald–Letnikov difference approximation and the algorithm has been examined both theoretically and experimentally.

Let us consider space-fractional convection–diffusion equation with variable coefficients:

$$\frac{\partial u(x,t)}{\partial t} + c(x)\frac{\partial u(x,t)}{\partial x} = d(x)\frac{\partial^\alpha u(x,t)}{\partial x^\alpha} + p(x,t), \quad x \in (L,R), \ t \in (0,T], \alpha \in (1,2]; \quad (1)$$

with the given initial condition:

$$u(x,0) = g(x), \quad L \leq x \leq R,$$

and homogeneous Dirichlet boundary conditions:

$$u(L,t) = 0, \ u(R,t) = 0, \ 0 \leq t \leq T,$$

where $c(x), d(x)$ and $g(x)$ are continuous functions on $[L,R]$ and $p(x,t)$ is continuous function on $[L,R] \times [0,T]$. Here $u(x,t)$ is the concentration, $d(x) > 0$ is the variable diffusion coefficient, $c(x) > 0$ is the fluid variable velocity which means the system is evolving in space due to a velocity field and $p(x,t)$ is sink term so that the fluid transport is from left to right. For the case of integer order ($\alpha = 2$), Equation (1) gives to the classical convection–diffusion equation(CDE). In this study, we have only considered the fractional derivative case which describes a physical meaning in [36] and it involves only a left-sided fractional order derivative. We have assumed that this one-dimensional space fractional convection–diffusion problem has sufficiently smooth and unique enough solutions.

The structure of this paper is arranged as follows. In Section 2, we introduce some preliminary remarks, lemmas and definitions and we show the formulation of the new Crank–Nicolson with right Shifted Grünwald–Letnikov difference scheme in Section 3. In Section 4, we describe the unconditional stability using *Gerschgorin* Theorem and convergence order analysis of the scheme. In Section 5, numerical tests are implemented to show the relevance of our theoretical study and the conclusions are put in Section 6.

2. Preliminary Remarks

Definition 1. *The Riemann fractional derivative operator D_*^α with order α is written as:*

$$(D_*^\alpha u)(x) = \frac{1}{\Gamma(r-\alpha)} \frac{d^r}{dx^r} \int_L^x \frac{u(t)}{(x-t)^{\alpha-r+1}} dt, \quad \alpha > 0 \tag{2}$$

where $r-1 < \alpha < r$, $r \in N$, $t > 0$.

Definition 2. *The left hand side and the right hand side fractional order derivatives, respectively, in Equation (1) are the Riemann–Liouville fractional derivatives with order α which are given by:*

$$\begin{aligned}
(D_+^\alpha u)(x) &= \frac{1}{\Gamma(r-\alpha)} \frac{d^r}{dx^r} \int_L^x (x-s)^{r-\alpha-1} u(s) ds \\
(D_-^\alpha u)(x) &= \frac{(-1)^r}{\Gamma(r-\alpha)} \frac{d^r}{dx^r} \int_x^R (s-x)^{r-\alpha-1} u(s) ds
\end{aligned} \tag{3}$$

for $r-1 < \alpha < r$, $x \in \Re$.

Definition 3 ([3]). *Let u be given on $\Re$. The standard Grünwald–Letnikov estimate for $1 < \alpha \leq 2$ with positive order α is defined by the formula,*

$$D^\alpha u(x,t) \approx \frac{1}{h^\alpha} \sum_{k=0}^{N_x} \omega_k^{(\alpha)} u(x-kh, t), \tag{4}$$

we also define the Grünwald–Letnikov difference operator as:

$$h^{-\alpha}(\Delta_h^\alpha u)(x,t) \approx \sum_{k=0}^{N_x} \omega_k^{(\alpha)} u(x-kh, t), h > 0, x \in \Re, \tag{5}$$

where

$$\omega_k^{(\alpha)} = \frac{\alpha(\alpha-1)...(\alpha-k+1)}{k!}, \tag{6}$$

is called Grünwald–Letnikov coefficient which is the Taylor series expansion $\omega(z) = (1-z)^\alpha$ which is the generating function. We can expressed the coefficients by the following recursive relations.

$$\omega_0^{(\alpha)} = 1, \omega_k^{(\alpha)} = (1 - \frac{\alpha+1}{k}) \omega_{k-1}^{(\alpha)}, k = 1, 2, ... \tag{7}$$

Lemma 1 ([37]). *Assume that $1 < \alpha \leq 2$, then Grünwald–Letnikov coefficients $\omega_k^{(\alpha)}$ satisfy:*

$$\begin{cases} \omega_0^{(\alpha)} = 1, \omega_1^{(\alpha)} = -\alpha < 0, \omega_2^{(\alpha)} = \frac{\alpha(\alpha-1)}{2} > 0 \\ 1 \geq \omega_2^{(\alpha)} \geq \omega_3^{(\alpha)} \geq ... \geq 0, \\ \sum_{k=0}^\infty \omega_k^{(\alpha)} = 0, \sum_{k=0}^{N_x} \omega_k^{(\alpha)} < 0, \ N_x \geq 1. \end{cases} \tag{8}$$

The Shifted Grünwald–Letnikov difference operator expression is suitable for our purpose because, it allows us to estimate $(D_*^\alpha u)(x)$, which is defined in Equation (2), numerically in an accurate way. According to [14], right shifted Grünwald–Letnikov difference operator with p shifts for α^{th} order Left R-L fractional derivative of $u(x,t), x \in [L,R]$ at $x = x_m$ can be expressed as:

$$(D_*^\alpha u)(x,t) \approx \frac{1}{h^\alpha} \sum_{k=0}^{\frac{x_m-L}{h}+p} \omega_k^{(\alpha)} u(x-(k-p)h, t) \tag{9}$$

where
$$x_m = L + mh, h = \frac{R-L}{N_x}, m = 0, 1, 2, \ldots N_x.$$

Lemma 2 ([38,39]). *Let $u \in C^{2n}(\Re)$ that has a finite degree of smoothness with $(D_+^\alpha u)(x)$ which is approximated by $h^{-\alpha}(\Delta_h^\alpha u)(x)$ possesses an asymptotic expansion in integer powers of the step-length h, then an expansion in even powers of h for the Shifted operator can be written in the form:*

$$\left(\Delta_{h,p}^\alpha u\right)(x) = \sum_{j=0}^{\infty} (-1)^j \binom{\alpha}{j} u\left(x + \frac{\alpha h}{2} - jh\right), h > 0. \tag{10}$$

Lemma 3 ([39]). *Let $u \in C^{n+3}(\Re)$ all derivative of u up to the order $n+4$ belong to $L^1(\Re)$. Then the Fourier transform of the Grünwald–Letnikov difference operator defined in Equation (5), is*

$$\hat{\phi}(x) = \int_\Re \phi(t) e^{ixt} dt. \tag{11}$$

Theorem 1. *Let $u \in C^{2n+3}(\Re)$ with all derivatives of u up to order $2n+3$ belong to $L^1(\Re)$. For $p \geq 0$ define the shifted Grünwald–Letnikov operator:*

$$(\Delta_{h,p}^\alpha) u(x) = \sum_{k=0}^{\infty} \omega_k^{(\alpha)} u\left(x - (k-p)h\right),$$

with $\omega_k^{(\alpha)} = (-1)^{2k} \binom{\alpha}{a_{2k}} = \binom{\alpha}{a_{2k}}$. Then, if $L = -\infty$ in Equation (2), for any computable coefficient a_{2k}, which is independent of h, u and x, we have

$$h^{-\alpha}\left(\Delta_{h,p}^\alpha u\right)(x) = (D_+^\alpha u)(x) + \sum_{k=1}^{n-1} b_{2k} \left(D_+^{\alpha+2k} u\right)(x) h^{2k} + O(h^{2n})$$

uniformly in $x \in \Re$.

Proof of Theorem 1. We closely follow the result described in [9,10] for the unshifted Grünwald–Letnikov formula and also in [23] for the shifted Grünwald–Letnikov formula. We can see that with the Riemann-Lebesgue lemma, the assumptions on u indicates for real positive constant C_1 and from the condition which is imposed on u, we have

$$|\tilde{u}(t)| \leq C_1 (1+|t|)^{-2n-3}. \tag{12}$$

From Lemma 3 for all $t \in \Re$ the Fourier transform for $u(x)$ of the Grünwald–Letnikov approximation is

$$\tilde{u}(t) = \int_\Re u(x) e^{ixt} dx.$$

From the definition of Fourier transform, we have observed that for a constant $a \in \Re$, we have:

$$\mathcal{F}([u(x-a)])(t) = e^{iat} \tilde{u}(t).$$

The function
$$\left(\frac{1-e^{-z}}{z}\right)^\alpha e^{zp} = \omega_{\alpha,p}(z),$$

have the Taylor expansion
$$\omega_{\alpha,p}(z) = \sum_{k=0}^{\infty} a_{2k} z^{2k}, \tag{13}$$

where $a_{2k}=(-1)^{2k}\binom{\alpha}{a_{2k}}=\binom{\alpha}{a_{2k}}$, converges absolutely for $|z| \leq 1$ since the function $\omega_{\alpha,p}(z)$ is bounded on $\Re$. The shifted Grünwald difference approximation $(\Delta_{h,p})u(x) \in L^1(\Re)$.

Thus, we have

$$\begin{aligned}\mathcal{F}(h^{-\alpha}\Delta_{h,p}^{\alpha}u)(t) &= h^{-\alpha}e^{-itph}\sum_{k=0}^{\infty}\binom{\alpha}{2k}e^{ikth}\tilde{u}(t)\\ &= h^{-\alpha}e^{-itph}\left(1-e^{itph}\right)^{\alpha}\tilde{u}(t)\\ &= (-it)^{\alpha}\left(\frac{1-e^{ith}}{-ith}\right)^{\alpha}e^{-itph}\tilde{u}(t) = (-it)^{\alpha}\omega_{\alpha,p}(-ith)\tilde{u}(t)\end{aligned} \qquad (14)$$

since $\omega_{\alpha,p}(-ith)$ is analytic around the origin, we express it as an even power expansions

$$\omega_{\alpha,p}(z) = \sum_{k=0}^{\infty} a_{2k}z^{2k}$$

which absolutely convergent for all $|z| \leq R$. For this a bounded function $\omega_{\alpha,p}(z)$ on $\Re$, there exist a real positive constant C_2 which satisfy:

$$\left|\left(\frac{1-e^{ix}}{-ix}\right)^{\alpha} - \sum_{k=0}^{n-1} a_{2k}(-ix)^{2k}\right| \leq C_2|x|^{2n} \qquad (15)$$

is bounded uniformly in $x \in \Re$. For any value $|x| \leq R$, we have

$$\left|(\omega_{\alpha,p}(-ix) - \sum_{k=0}^{n-1} a_{2k}(-ix)^{2k}\right| = \left|\sum_{k=n}^{\infty} a_{2k}(-ix)^{2k}\right| \leq |x|^{2n}\sum_{k=n}^{\infty}\binom{\alpha}{a_{2k}}|x|^{2(k-n)} \leq C_3|x|^{2n} \qquad (16)$$

which is bounded on $\Re$. For the other case $|x| > R$ also, we have

$$|\omega_{\alpha,p}(-ix)| = \left|\left(\frac{1-e^{ix}}{-ix}\right)^{\alpha}e^{ipx}\right| \leq \frac{2^{\alpha}}{R^{\alpha}} < C_4|x|^{2n} \qquad (17)$$

where $C_4 = \frac{2^{\alpha}}{R^{\alpha+2n}} < \infty$ and also

$$\left|\sum_{k=0}^{n-1} a_{2k}(-ix)^{2k}\right| \leq |x|^{2n}\sum_{k=0}^{n-1}\left|\binom{\alpha}{a_{2k}}\right||x|^{2(k-n)} \leq C_5|x|^{2n} \qquad (18)$$

with $C_5 = \sum_{k=0}^{n-1}\left|\binom{\alpha}{a_{2k}}\right|R^{2k-2n} < \infty$. Now, we set that

$$C_2 = \max\left\{\sum_{k=0}^{\infty}|a_{2k}|R^{2k-2n}, \frac{2^{\alpha}}{R^{\alpha+2n}} + \sum_{k=0}^{n-1}|a_{2k}|R^{2k-2n}\right\}$$

since

$$\sum_{k=0}^{\infty}|a_{2k}|R^{2k-2n} = \sum_{k=0}^{n-1}|a_{2k}|R^{2k-2n} + \sum_{k=n}^{\infty}|a_{2k}|R^{2k-2n}$$

$$C_2 = \frac{2^{\alpha}}{R^{\alpha+2n}} + \sum_{k=0}^{n-1}|a_{2k}|R^{2k-2n}$$

Then, this implies that Equation (15) holds for all $x \in \Re$. From Equation (17), we can write

$$\mathcal{F}(h^{-\alpha}\Delta^{\alpha}_{h,p}u)(t) = \sum_{k=0}^{n-1} a_{2k}(-it)^{\alpha+2k} h^{2k}\tilde{u}(t) + \tilde{\varphi}(t,h)$$

where

$$\tilde{\varphi}(k,h) = (-it)^{\alpha}\left(\omega_{\alpha,p}(-ith) - \sum_{k=0}^{n-1} a_{2k}(-ith)^{2k}\right)\tilde{u}(t)$$

since

$$(-it)^{\alpha+2k}\tilde{u}(t) = \left(D^{\alpha+2k}_{+}\right)\tilde{u}(t).$$

Therefore, we have

$$(-it)^{\alpha+2k}\tilde{u}(t) \in L^1(\Re).$$

Moreover, we see that

$$\tilde{\varphi}(t,h) \in L^1(\Re),$$

and with the conditions imposed on u, we can say that $(1+|x|^{2n+3})\tilde{u}(t)$ is bounded on $\Re$. Thus, $|t|^{2\alpha-3}|\tilde{u}(t)| \in L^1(R)$. This implies that,

$$|\tilde{\varphi}(t,h)| \leq Ch^{2n}(1+|t|)^{2\alpha-3}$$

for $k \in \Re$ with $C = C_1 C_2$. Therefore using the Fourier inversion transform, we have

$$h^{-\alpha}\left(\Delta^{\alpha}_{h,p}u\right)(x) = (D^{\alpha}_{+}u)(x) + \sum_{k=1}^{n-1} a_{2k}\left(D^{\alpha+2k}_{+}u\right)(x)h^{2k} + \varphi(x,h),$$

where

$$\varphi(x,h) = \left|C\int_{R} e^{-itx}\tilde{\varphi}(t,h)dt\right| \leq C\int_{R}|\tilde{\varphi}(t,h)dt| \leq Ch^{2n}.$$

At last, we have

$$h^{-\alpha}\left(\Delta^{\alpha}_{h,p}u\right)(x) = (D^{\alpha}_{+}u)(x) + \sum_{k=1}^{n-1} a_{2k}(D^{\alpha+2k}_{+}u)(x)h^{2k} + O(h^{2n}). \tag{19}$$

☐

Remark 1. *From Equation (10), it can be seen that for $p = \alpha/2$, the error takes its minimum value and a second order convergence is achieved. We need the grid points $x_m - (k-p)h$ to find an optimal positive integer p that makes $p - \alpha/2$ is minimum. It is numerically proved in [3] that for the value $0 < \alpha \leq 1, p = 0$ is acceptable; while for $1 < \alpha \leq 2, p = 1$ is optimal.*

Remark 2. *Theorem 1 is the base of Extrapolation to the limit. Therefore one can apply it the Shifted Grünwald–Letnikov difference operator to obtain the convergence rate with arbitrary high order $h^k, k = 1, 2, 3, ..., n$ such that*

$$h^{-\alpha}\frac{(q^{-\alpha}\Delta^{\alpha}_{qh,p}u)(x) - q(\Delta^{\alpha}_{h,p}u)(x)}{1-q}, 0 < q < 1$$

(q is fixed) converges to $(D^{\alpha}_{+}u)(x) + O(h^2)$.

3. Problem Formulation of the Scheme

Consider the following one-dimensional space fractional convection–diffusion problem:

$$\begin{cases} \dfrac{\partial u(x,t)}{\partial t} = -c(x)\dfrac{\partial u(x,t)}{\partial x} + d(x)\dfrac{\partial^\alpha u(x,t)}{\partial x^\alpha} + p(x,t), & (x,t) \in (L,R) \times (0,T] \\ u(x,0) = g(x), & x \in [L,R] \\ u(L,t) = 0, u(R,t) = 0, & t \in [0,T] \end{cases} \quad (20)$$

which is based on shifted Grünwald–Letnikov difference method with $1 < \alpha \leq 2$ on a finite domain $L < x < R$.

Crank–Nicolson Scheme for Time and Shifted Grünwald Difference Scheme for Space Discretization

We partition the finite interval $[L,R]$ with a uniform mesh in the space size step $h = (R-L)/N_x$ and the time step $\tau = T/N_t$, in which N_x, N_t are non-negative integers and the set of grid size points is symbolized by $x_m = mh$ and $t_n = n\tau$ for $0 \leq m \leq N_x$, $0 \leq n \leq N_t$. Set $t_{n+1/2} = (t_{n+1} + t_n)/2$ with $0 \leq n \leq N_t - 1$.

We use the following notations:

$u_m^n = u(x_m, t_n), p_m^{n+1/2} = p(x_m, t_{n+1/2}), \delta_t u_m^n = \dfrac{u_m^{n+1} - u_m^n}{\tau}, c_m = c(x_m), d_m = d(x_m).$

Applying the C-N technique for the time discretization of Equation (20) gives to

$$\begin{aligned} \delta_t u_m^n &= -\dfrac{c_m}{4h}\left(u_{m+1}^{n+1} - u_{m-1}^{n+1} + u_{m+1}^n - u_{m-1}^n\right) \\ &+ \dfrac{d_m}{2h^\alpha}\sum_{z=0}^{1}\sum_{k=0}^{N_x-1}\omega_k^{(\alpha)}\left(u_{m-k+1}^{n+z}\right) = p_m^{n+1/2} + O(\tau^2). \end{aligned} \quad (21)$$

In space discretization we have used the central finite difference method for the convection term and the Shifted Grünwald–Letnikov operator for the space fractional derivative with the approach of spatial Extrapolation to the limit, respectively.

See the full discretization of the scheme:

$$\begin{aligned} \dfrac{u_m^{n+1} - u_m^n}{\tau} &= \dfrac{-c_m\left(u_{m+1}^n - u_{m-1}^n + u_{m+1}^{n+1} - u_{m-1}^{n+1}\right)}{4h} \\ &+ \dfrac{d_m}{2h^\alpha}\left(\sum_{z=0}^{1}\sum_{k=0}^{m+1}\omega_k^{(\alpha)}u_{m-k+1}^{n+z}\right) + \dfrac{p_m^n + p_m^{n+1}}{2}. \end{aligned} \quad (22)$$

Multiplying Equation (22) by τ the discretization equation, we have

$$\begin{aligned} u_m^{n+1} - u_m^n &= \dfrac{-c_m \tau}{4h}(u_{m+1}^n - u_{m-1}^n + u_{m+1}^{n+1} - u_{m-1}^{n+1}) + \\ & \dfrac{d_m \tau}{2h^\alpha}\sum_{z=0}^{1}\sum_{k=0}^{m+1}\omega_k^{(\alpha)}u_{m-k+1}^{n+z} + \tau p_m^{n+1/2} \end{aligned} \quad (23)$$

The above equation is used to predict the values of $u(x,t)$ at time $n+1$, so all the values of u at time n are assumed to be known. For simplification

$\mu_m = \frac{c_m \tau}{h}, \eta_m = \frac{d_m \tau}{h^\alpha}$, then we have

$$\left(1 - \frac{\eta_m}{2}\omega_1^\alpha\right) u_m^{n+1} + \left(\frac{-\mu_m}{2} - \frac{\eta_m}{2}\omega_2^\alpha\right) u_{m-1}^{n+1}$$
$$+ \left(-\frac{\mu_m}{2} - \frac{\eta_m}{2}\omega_0^\alpha\right) u_{m+1}^{n+1} - \frac{\eta_m}{2}\left(\sum_{k=3}^{m+1} \omega_k^\alpha u_{m-k+1}^{n+1}\right)$$
$$= \left(1 + \frac{\eta_m}{2}\omega_1^\alpha\right) u_m^n + \left(\frac{\mu_m}{2} + \frac{\eta_m}{2}\omega_2^\alpha\right) u_{m-1}^n$$
$$+ \left(\frac{\eta_m}{2}\omega_0^\alpha + \frac{\mu_m}{2}\right) u_{m+1}^n + \frac{\eta_m}{2}\left(\sum_{k=3}^{m+1} \omega_k^{(\alpha)} u_{m-k+1}^n\right) + \tau\left(p_m^{n+\frac{1}{2}}\right). \quad (24)$$

Both the convection and diffusion variable coefficients are $(N_x - 1) \times (N_x - 1)$ diagonal matrices which are defined by

$$\mu_m = \frac{\tau}{2h} \operatorname{diag}(C_1, C_2, C_3, \ldots C_{N_x-1}),$$
$$\eta_m = \frac{\tau}{h^\alpha} \operatorname{diag}(d_1, d_2, d_3, \ldots d_{N_x-1}).$$

These discretization together with Dirichlet boundary conditions which results in a linear system of equations for which the coefficient matrix is the sum of lower triangular and upper-diagonal matrices. The above discretization can be re-arranged to yield:

$$\left(1 - \frac{\eta_m}{2}\omega_1^\alpha\right) u_m^{n+1} + \left(-\frac{\mu_m}{2} - \frac{\eta_m}{2}\omega_2^\alpha\right) u_{m-1}^{n+1} +$$
$$\left(-\frac{\mu_m}{2} - \frac{\eta_m}{2}\omega_0^\alpha\right) u_{m+1}^{n+1} - \frac{\eta_m}{2}\left(\sum_{k=3}^{m+1} \omega_k^\alpha u_{m-k+1}^{n+1}\right)$$
$$= (1 + \frac{\eta_m}{2}\omega_1^\alpha) u_m^n + \left(\frac{\mu_m}{2} + \frac{\eta_m}{2}\omega_2^\alpha\right) u_{m-1}^n$$
$$+ (\frac{\eta_m}{2}\omega_0^\alpha + \frac{\mu_m}{2}) u_{m+1}^n + \frac{\eta_m}{2}\left(\sum_{k=3}^{m+1} \omega_k^\alpha u_{m-k+1}^n\right) + \tau(P_m^{n+\frac{1}{2}}). \quad (25)$$

Denoting U_m^n as the numerical approximation of u_m^n, we can construct the C-N scheme for Equation (20)

$$\left(1 - \frac{\eta_m}{2}\omega_1^\alpha\right) U_m^{n+1} + \left(-\frac{\mu_m}{2} - \frac{\eta_m}{2}\omega_2^\alpha\right) U_{m-1}^{n+1} +$$
$$\left(-\frac{\mu_m}{2} - \frac{\eta_m}{2}\omega_0^\alpha\right) U_{m+1}^{n+1} - \frac{\eta_m}{2}\left(\sum_{k=3}^{m+1} \omega_k^\alpha U_{m-k+1}^{n+1}\right)$$
$$= \left(1 + \frac{\eta_m}{2}\omega_1^\alpha\right) U_m^n + \left(\frac{\mu_m}{2} + \frac{\eta_m}{2}\omega_2^\alpha\right) U_{m-1}^n$$
$$+ \left(\frac{\eta_m}{2}\omega_0^\alpha + \frac{\mu_m}{2}\right) U_{m+1}^n + \frac{\eta_m}{2}\left(\sum_{k=3}^{m+1} \omega_k^\alpha U_{m-k+1}^n\right) + \tau(P_m^{n+\frac{1}{2}}). \quad (26)$$

I is the $(N_x - 1) \times (N_t - 1)$ identity matrix with $A_{m,n}$ as the matrix coefficients. These coefficients, for $m = 1, 2, 3, \ldots, N_x - 1, n = 1, 2, \ldots, N_t - 1$ are given by:

$$A_{m,n} = \begin{cases} 0, & n \geq m+2 \\ -\frac{\mu_m}{2} - \frac{\eta_m}{2}\omega_0^{(\alpha)}, & n = m+1 \\ (1 - \frac{\eta_m}{2}\omega_1^{(\alpha)}), & n = m \\ (-\frac{\eta_m}{2}\omega_2^{(\alpha)} - \frac{\mu_m}{2}), & n = m-1 \\ -\frac{\eta_m}{2}\omega_{m-n+1}^{(\alpha)} & n \leq m-1. \end{cases} \quad (27)$$

The finite difference scheme (24) and (26) defines a linear system of equations as

$$(I+A)U^{n+1} = (I-A)U^n + \tau(p_m^{n+\frac{1}{2}}) \tag{28}$$
$$U^{n+1} = [u_1^{n+1}, u_2^{n+1}, ..., u_{N_x-1}^{n+1}]^\top$$
$$U^n + \tau P_m^{n+\frac{1}{2}} = [0, \tau p_1^{n+\frac{1}{2}}, \tau p_2^{n+\frac{1}{2}}, ..., \tau p_{N_x-1}^{n+\frac{1}{2}} + (\frac{\eta_{N_x-1}}{2} + \frac{\mu_{N_x-1}}{2}), 0]^\top.$$

Theorem 2. *Suppose that $1 < \alpha \leq 2$, the coefficient matrix defined in Equations (24)–(27), then the diagonal matrix and the coefficient matrix satisfy:*

$$A_{m,m} > \sum_{n=0, m\neq 1}^{N_x-1} |A_{m,n}|, m = 1, 2, 3, ..., N_x - 1. \tag{29}$$

Proof of Theorem 2. As we have seen from the coefficient matrix defined in Equation (27),
$A_{m,m+1} = \frac{\mu_m}{2} - \frac{\eta_m}{2}\omega_0^{(\alpha)} = \frac{\mu_m}{2} - \frac{\eta_m}{2} < 0$
$A_{m,m-1} = -\frac{\eta_m}{2}\omega_2^{(\alpha)} - \frac{\mu_m}{2} = -\frac{\eta_m}{2}(\frac{\alpha^2-\alpha}{2})$, but from Lemma 1, $\frac{\alpha^2-\alpha}{2} > 0$ for $1 < \alpha \leq 2$ mean that
$-\frac{\eta_m}{2}(\frac{\alpha^2-\alpha}{2}) < 0$.

When $n < m-1$, we have, $-\frac{\eta_m}{2}\omega_{m-n+1}^{(\alpha)} < 0$ and when $n = m$, $A_{m,m} = 1 - \frac{\eta_m}{2}\omega_1^{(\alpha)} = 1 + \frac{\eta_m}{2}\alpha > 0$.
This implies that $\sum_{n=0, m\neq 1}^{N_x-1} |A_{m,n}| < A_{m,m}$.
Therefore, the diagonal matrix is strictly dominant. □

4. Theoretical Analysis of Finite Difference Scheme

In general for analyzing convergence and stability, we consider the following description.
Let $\chi_h = \{v : v = \{v_m\} : \{x_m = mh\}_{m=0}^{N_x}, v_0 = v_{N_x} = 0\}$ be the grid function.
For any $v = v_m \in \chi_h$, we define our point-wise maximum norm as

$$||v||_\infty = \max_{1 \leq m \leq N_x} |v_m|, \tag{30}$$

and the discrete L^2-norm

$$||v|| = \sqrt{h \sum_{m=1}^{N_x-1} v_m^2}. \tag{31}$$

4.1. Boundedness of the Fractional Scheme

The Classical Crank–Nicolson scheme combines the stability of an implicit finite difference method with its accuracy which produce second order convergence in both space and time.

Theorem 3. *Crank–Nicolson scheme for solving space fractional convection–diffusion equations given by the following problem:*

$$\frac{\partial u(x,t)}{\partial t} + c(x)\frac{\partial u(x,t)}{\partial x} = d(x)\frac{\partial^\alpha u(x,t)}{\partial x^\alpha} + p(x,t). \tag{32}$$

which is based on shifted Grünwald–Letnikov difference approximation scheme is bounded for $1 < \alpha \leq 2$.

Proof of Theorem 3. Consider C-N scheme for the space-fractional convection–diffusion problem for $1 < \alpha \leq 2$

$$\frac{u_m^{n+1} - u_m^n}{\tau} = \frac{-c_m \left(u_m^n - u_{m-1}^n + u_m^{n+1} - u_{m-1}^{n+1}\right)}{2h}$$
$$+ \frac{d_m}{2h^\alpha} \left(\sum_{j=0}^{1} \sum_{k=0}^{m+1} \omega_k^{(\alpha)} u_{m-k+1}^{n+j}\right) + \frac{p_m^n + p_m^{n+1}}{2}. \tag{33}$$

Here, we have shown the convergence and boundedness of the scheme by taking the smaller time-step in terms of Lax–Richtmyer stability analysis that uses a weaker bound (see [40]). Our matrix A has an eigenvalues of λ that have positive real parts, and, we also have found a strictly dominant matrix. These eigenvalues which are centered in the disks at each diagonal entries as:

$$A_{m,m} = \left(1 - \frac{\eta_m}{2}\omega_1^\alpha\right) = \left(1 + \alpha \frac{\eta_m}{2}\right).$$

with $\mu_m = \frac{c_m \tau}{h}, \eta_m = \frac{\tau d_m}{h^\alpha}$. From the *Gerschgorin* Theorem in [41], the radius of the matrix can be expressed as

$$\left\|\sum_{n=0, m \neq 1}^{N_x} A_{m,n}\right\|_2^2 = \left\|\left(-\frac{\eta_m}{2} - \frac{\mu_m}{2}\right) \sum_{n=0}^{m+1} \omega_{m-n+1}^{(\alpha)}\right\|_2^2$$
$$\leq \left\|\left(-\frac{\eta_m}{2} - \frac{\mu_m}{2}\right)\right\|^2 \left\|\sum_{n=0}^{m+1} \omega_{m-n+1}^{(\alpha)}\right\|^2.$$

Since from the Grünwald coefficients we have $\omega_{m-n+1}^{(\alpha)} \leq \omega_1^{(\alpha)}$ and $\omega_1^{(\alpha)} = -\alpha$, we have that:

$$\left\|\sum_{n=0, m \neq 1}^{N_x} (A_{m,n})\right\|_2^2 \leq \left|\sum_{n=0, m \neq 1}^{N_x} (A_{m,m})\right|_2^2 \leq \|A_{m,m}\|_2^2$$
$$\leq \left\|\left(-\frac{\eta_m}{2} - \frac{\mu_m}{2}\right)\right\|_2^2 \|\omega_1^{(\alpha)}\|_2^2 \leq \left\|1 + \frac{\eta_m}{2}\alpha\right\|_2^2.$$

For a bounded ratio of time-step τ and space-step h with $n\tau \leq T$, we have

$$\|(A_{m,m})^n\|_2 \leq \left(1 + \frac{\eta_m}{2}\alpha\right)^{n/2}.$$

From the relation of Parseval's Theorem, [40]

$$\|A_{m,m}\|_2 \leq \left(1 + \frac{\eta_m}{2}\alpha\right)^{n/2} \leq e^{\alpha T/2}.$$

which shows that the scheme is bounded. □

4.2. Stability Analysis

Theorem 4. *Let U_m^n be the numerical approximation of the exact solution u_m^n, then the C-N finite difference scheme (28) is unconditionally stable.*

Proof of Theorem 4. Consider the matrix coefficient of the difference approximation for the problem (20) can be written as described above

$$(I + A)U^{n+1} = (I - A)U^n + \tau p_m^{n+1/2}. \tag{34}$$

Let $e^n = \{e_1^n, e_2^n, e_3^n, ..., e_{N_x-1}^n\}$, and take the relation between the error e^{n+1} in U^{n+1} and the error e^n in U^n which is given by the linear system

$$e^{n+1} = (I+A)^{-1}(I-A)e^n. \tag{35}$$

First of all, we must show that the (non-real valued) eigenvalues of the coefficient matrices A have positive real parts. For $w_1^{(\alpha)} = -\alpha$ with fractional order $1 < \alpha < 2$ and $k \neq 1$; we have $w_k^{(\alpha)} > 0$. In addition to this, $-w_1^\alpha = \alpha \geq \sum_{k=0, k\neq 1}^N w_k^\alpha$ for the value $N > 1$. As stated in *Gerschgorin* Theorem ([41], pp. 136–139), the eigenvalues of the given matrix A are inside the disks centered at each diagonal entry.

$$A_{m,m} = (1 - \frac{\eta_m}{2} w_1^{(\alpha)}) = 1 + \frac{\eta_m}{2}\alpha > 0,$$

with radius

$$r_m = \sum_{n=0, m\neq 1}^{N_x} |A_{m,n}| = \frac{\eta_m}{2} \sum_{n=0}^{m+1} w_{m-n+1}^{(\alpha)} < (1 + \frac{\eta_m}{2}).$$

These *Gerschgorin* disks are belong to the right half of the complex plane. Thus, the eigenvalue of the coefficient matrix A has positive real part which implies that A has an eigenvalue λ if and only if $(I-A)$ has an eigenvalue $(1-\lambda)$ if and only if $(I+A)^{-1}(I-A)$ has an eigenvalue $\left(\frac{1-\lambda}{1+\lambda}\right)$. From the first part of this sentence, we have seen that all the eigenvalues of the matrix given by $(I+A)$ have a radius larger than unity which implies the matrix is invertible. Now we can see from the above description the real part of λ is non-negative which we can conclude that $\left|\frac{(1-\lambda)}{(1+\lambda)}\right| < 1$. Thus, the spectral radius of the system matrix $(I+A)^{-1}(I-A)$ is strictly less than unity which implies that the difference scheme is unconditionally stable. □

4.3. Convergence Analysis

First of all we have given the Truncation error of the C-N scheme. It is obvious to conclude that:

$$\frac{u(x_m, t_{n+1}) - u(x_m, t_n)}{\tau} = \left(\frac{\partial u(x,t)}{\partial t}\right)^{n+1/2} + O(\tau^2).$$

$$\left(c(x)\frac{\partial u(x,t)}{\partial x} + d(x)\frac{\partial^\alpha u(x,t)}{\partial x^\alpha}\right)_m^{n+1/2} = \frac{1}{2}\left(c_m\frac{\partial u(x_m, t_{n+1})}{\partial x} + d_m\frac{\partial^\alpha u(x_m, t_{n+1})}{\partial x^\alpha}\right)$$
$$+ \frac{1}{2}\left(c_m\frac{\partial u(x_m, t_n)}{\partial x} + d_m\frac{\partial^\alpha u(x_m, t_n)}{\partial x^\alpha}\right) + O(\tau^2). \tag{36}$$

$$c(x_m)\frac{\partial u(x,t)}{\partial x} \approx \frac{u(x_{m+1}, t_{n+1}) - u(x_{m-1}, t_{n+1})}{2h} + O(h^2). \tag{37}$$

From the above Extrapolation to the limit Theorem for $n = 1$, we got

$$\frac{\partial^\alpha u(x,t)}{\partial x^\alpha} \approx \sum_{k=0}^{m+1} g_k^{(\alpha)} u_{m-k+1} + O(h^2). \tag{38}$$

Therefore the local truncation error of (20) is given by $T_m^{n+1} = O(\tau^2 + \tau h)$

Theorem 5. *Let u_m^n be the exact solution of problem (20), and U_m^n be the solution of the finite difference scheme (26), then for all $1 \leq n \leq N_t$, we have the estimate*

$$\|u_m^n - U_m^n\|_\infty \leq c(\tau^2 + h)$$

where $\|u_m^n - U_m^n\|_\infty = max_{1 \leq m \leq N_x} |u_m^n - U_m^n|$, c is a non-negative constant independent of h and τ with $\|.\|$ stands for the discrete L^2-norm.

Proof of Theorem 5. Denote $e^n = u_m^n - U_m^n$ where $e^n = (e_1^n, e_2^n, ..., e_{N_x-1}^n)$. We have $e^0 = 0$, we have from Equations (26) and (27) if $n = 0$,

$$R_m^1 = \left(\frac{-\mu_m}{2} - \frac{\eta_m}{2}\omega_0^{(\alpha)}\right) e_{m-1}^1 + \left(1 + \frac{\eta_m}{2}\alpha\right) e_1^m$$
$$+ \left(\frac{-\mu_m}{2} - \frac{\eta_m}{2}\omega_2^{(\alpha)}\right) e_{m+1}^1 - \frac{\eta_m}{2} \sum_{k=3}^{N_x} \omega_k^{(\alpha)} e_{m-n+1}^1.$$

if $n > 0$,

$$R_m^{n+1} = \left(\frac{-\mu_m}{2} - \frac{\eta_m}{2}\omega_0^{(\alpha)}\right) e_{m-1}^{n+1} + \left(1 + \frac{\eta_m}{2}\alpha\right) e_{n+1}^m$$
$$+ \left(\frac{-\mu_m}{2} - \frac{\eta_m}{2}\omega_2^{(\alpha)}\right) e_{m+1}^{n+1} - \frac{\eta_m}{2} \sum_{k=3}^{N_x} \omega_k^{(\alpha)} e_{m-n+1}^{n+1}.$$

where $R_m^{n+1} \leq c(\tau^2 + h)$, $m = 1, 2, ..., N_x - 1$, $n = 1, 2, 3, ..., N_t - 1$, c is non-negative constant independent of h and τ.

We can use the mathematical induction to prove the Theorem. Let $n = 1$ and assume $|e_j| = max_{1 \leq m \leq N_x - 1} |e_m^1|$, we have the following expression.

$$||e^1||_\infty = |e_j^1| \leq \left(\frac{-\mu_j}{2} - \frac{\eta_j}{2}\omega_0^{(\alpha)}\right) |e_{j-1}^1| + \left(1 + \frac{\eta_j}{2}\alpha\right) |e_1^j|$$
$$+ \left(\frac{-\mu_j}{2} - \frac{\eta_j}{2}\omega_2^{(\alpha)}\right) |e_{j+1}^1| - \frac{\eta_j}{2} \sum_{k=3}^{N_x} \omega_k^{(\alpha)} |e_{j-n+1}^1|$$
$$\leq \left|\left(\frac{-\mu_j}{2} - \frac{\eta_j}{2}\omega_0^{(\alpha)}\right) e_{j-1}^1 + \left(1 + \frac{\eta_j}{2}\alpha\right) e_1^j + \left(\frac{-\mu_j}{2} - \frac{\eta_j}{2}\omega_2^{(\alpha)}\right) e_{j+1}^1 - \frac{\eta_j}{2} \sum_{k=3}^{N_x} \omega_k^{(\alpha)} e_{j-n+1}^1\right|$$
$$= |R_j^1| \leq c(\tau^2 + h).$$

Suppose that if $n \leq r$, $||e^r||_\infty \leq c(\tau^2 + h^2)$ hold and assume $n = r + 1$, let $|e_j^{r+1}| = max_{1 \leq m \leq N_x - 1} |e_m^{r+1}|$, notice that from Lemma 1, we have $\sum_{k=0}^{N_x} \omega_k^{(\alpha)} < 0$, $m = 1, 2, ..., N_x$. Therefore,

$$||e^{r+1}||_\infty = |e_j^{r+1}| \leq \left(\frac{-\mu_j}{2} - \frac{\eta_j}{2}\omega_0^{(\alpha)}\right) |e_{j-1}^{r+1}| + \left(1 + \frac{\eta_j}{2}\alpha\right) |e_{r+1}^j|$$
$$+ \left(\frac{-\mu_j}{2} - \frac{\eta_j}{2}\omega_2^{(\alpha)}\right) |e_{j+1}^{r+1}| - \frac{\eta_j}{2} \sum_{k=3}^{N_x} \omega_k^{(\alpha)} |e_{j-n+1}^{r+1}|$$
$$\leq \left|\left(\frac{-\mu_j}{2} - \frac{\eta_j}{2}\omega_0^{(\alpha)}\right) e_{j-1}^{r+1} + \left(1 + \frac{\eta_j}{2}\alpha\right) e_{r+1}^j + \left(\frac{-\mu_j}{2} - \frac{\eta_j}{2}\omega_2^{(\alpha)}\right) e_{j+1}^{r+1} - \frac{\eta_j}{2} \sum_{k=3}^{N_x} \omega_k^{(\alpha)} e_{j-n+1}^{r+1}\right|$$
$$= |R_j^{r+1}| \leq c(\tau^2 + h).$$

which completes the proof. □

Remark 3. *The Crank–Nicolson scheme, for classical convection–diffusion equation, provides stable C-N finite difference method that is second order convergence in time and space. Also a study based on C-N finite difference method with the spatial extrapolation to the limit method, see Theorem 1, is used to get temporal and spatial second order for one-sided SFCDEs with space variable coefficients.*

5. Numerical Tests

Problem test 1

1. Consider the space-fractional diffusion type of problem:

$$\frac{\partial u(x,t)}{\partial t} = d(x)\frac{\partial^\alpha u(x,t)}{\partial x^\alpha} + p(x,t)$$

with initial condition

$$u(x,0) = (x^2 - x^3); 0 \leq x \leq 1$$

homogeneous Dirichlet boundary condition

$$u(0,t) = 0 = u(1,t)$$

with variable diffusion coefficient,

$$d(x) = \Gamma(1.2)x^\alpha,$$

and source term

$$p(x,t) = (6x^3 - 3x^2)e^{-t}$$

The exact solution is

$$u(x,t) = (x^2 - x^3)e^{-t}$$

All numerical experiments are implemented using Theorem 1 and C-N scheme with the space domain, $0 < x < 1$ and time domain, $0 < t < T$. Figure 1 shows the maximum error produced by C-N scheme for large enough time domain and numerical solution is close enough to the exact solution using C-N scheme with $\alpha = 1.5$ in Figure 2. The maximum error and second order convergence for the fractional diffusion and fractional convection–diffusion equation with variable coefficients are given in Tables 1–3.

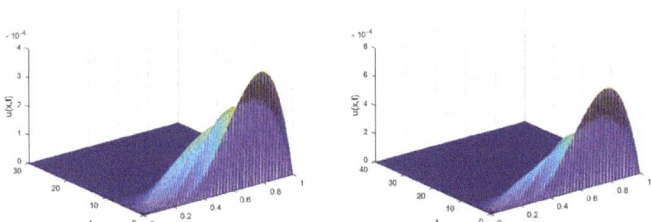

Figure 1. The Maximum error by C-N scheme at $(T = 10, Max - Error = 6.5276e^{-07})$, $(T = 20, Max - Error = 1.7244e^{-08})$, $\alpha = 1.5$ left to right, respectively, for example 1.

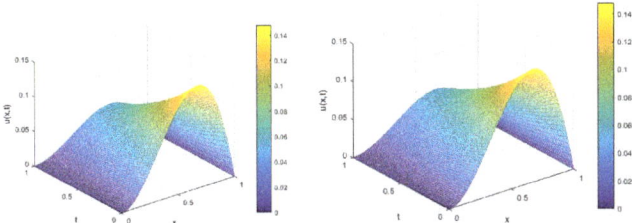

Figure 2. The exact (**left**) and numerical (**right**) solution by C-N scheme at $T = 1, \alpha = 1.5, \tau = 0.01 = h$ for example 1.

Table 1. The maximum error and convergence order of the C-N scheme for FDE in example 1.

		$\alpha = 1.25$		$\alpha = 1.5$		$\alpha = 1.8$	
Δt	Δx	Max-Error	Order	Max-Error	Order	Max-Error	Order
1/50	1/50	4.9807e−04	–	4.0046e−04	–	1.4048e−04	–
1/100	1/100	1.0660e−04	2.2241	8.8946e−05	2.1707	3.6848e−05	1.9307
1/200	1/200	2.4413e−05	2.1265	2.0643e−05	2.1073	9.4393e−06	1.9648
1/400	1/400	5.8239e−06	2.0676	4.9592e−06	2.0575	2.3887e−06	1.9825
1/800	1/800	1.4211e−06	2.0350	1.2146e−06	2.0296	6.0078e−07	1.9913

Table 2. The maximum error and convergence order for FCDE in example 2.

		$T = 1$		$T = 5$	
Δt	Δx	Max-Error	Order	Max-Error	Order
1/50	1/50	1.4048e−04	–	2.5297e−05	–
1/100	1/100	3.6848e−05	1.9307	7.4748e−06	1.7589
1/200	1/200	9.4393e−06	1.9648	2.0122e−06	1.8933
1/400	1/400	2.3887e−06	1.9825	4.9017e−07	2.0374
1/800	1/800	6.0078e−07	1.9913	1.0620e−07	2.2065

Table 3. The maximum error and convergence order by C-N for SFCDE in example 2 at $T = 1, \alpha = 1.55$.

Δt	Δx	Max-Error	Order
1/50	1/50	2.6e−03	–
1/100	1/100	7.695e−04	1.7563
1/150	1/150	2.144e−04	1.8436
1/200	1/200	5.688e−05	1.9143

Problem test 2

2. Consider the space-fractional convection–diffusion type of equation with variable coefficients:

$$\frac{\partial u(x,t)}{\partial t} + c(x)\frac{\partial u(x,t)}{\partial x} = d(x)\frac{\partial^\alpha u(x,t)}{\partial x^\alpha} + p(x,t)$$

with initial condition

$$u(x,0) = (x^\alpha - x); 0 \leq x \leq 1$$

homogeneous Dirichlet boundary condition

$$u(0,t) = 0 = u(1,t)$$

with variable convection–diffusion coefficients respectively,

$$c(x) = x^{\frac{1}{5}}, d(x) = x^{\frac{1}{100}},$$

and source term

$$p(x,t) = e^{-2t}\left(2(x - x^\alpha) - \Gamma(\alpha) + \frac{\Gamma(\alpha+1)}{\Gamma(\alpha)}x^{\alpha-1} - 1\right)$$

The exact solution is

$$u(x,t) = e^{-2t}(x^\alpha - x)$$

Figures 3 and 4 show the numerical and exact solutions for fractional diffusion and fractional convection–diffusion problems with large enough time domain in example 1 and 2, respectively. The exact and numerical solution of fractional convection–diffusion equation by C-N scheme is also given in Figure 5. In Table 4, the maximum error and first order convergence in space is obtained using C-N scheme without extrapolation to the limit approach by fixing the time step.

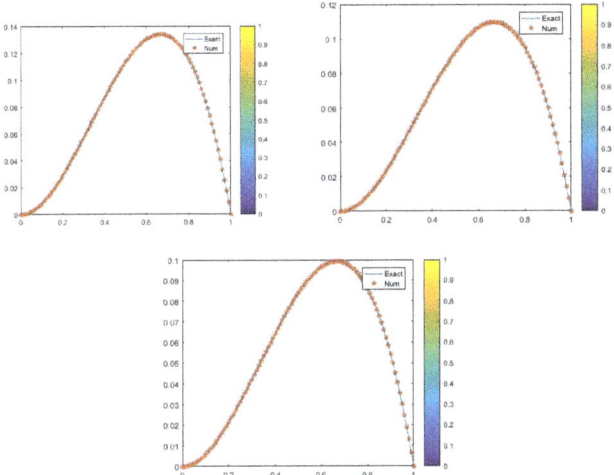

Figure 3. Numerical and exact solution by C-N scheme at $\alpha = 1.5, \tau = h = 0.01$, with ($T = 10, T = 30, T = 40$) left to right-down respectively, for example 1.

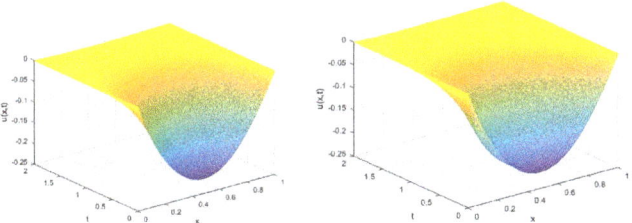

Figure 4. The exact (**left**) and numerical (**right**) solution by C-N scheme for the FCDE at ($h = \tau = 0.005, \alpha = 1.5, (t = 5, \max - error = 4.0657e^{-05})$ for example 2.

Table 4. The Maximum error and convergence order produced by C-N scheme for example 3 at $T = 1, N_t = 100$.

	$\alpha = 1.35$		$\alpha = 1.5$		$\alpha = 1.75$	
Δx	Max-Error	Order	Max-Error	Order	Max-Error	Order
1/50	4.5e−03	–	2.8e−03	–	1.7e−03	–
1/100	2.7e−03	0.7370	1.6e−03	0.8074	8.9641−04	0.97224
1/200	1.6e−03	0.7549	8.6405e−04	0.8889	4.6491e−04	0.8981
1/400	9.5896e−04	0.7385	4.7955e−04	0.8494	2.4086e−04	0.9488
1/800	5.7034e−04	0.7496	2.6609e−04	0.8498	1.2473e−04	0.9494

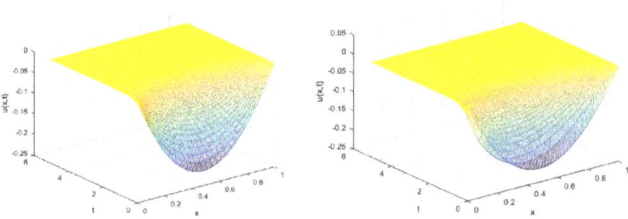

Figure 5. The exact (**left**) and numerical (**right**) solution by C-N scheme for the FCDE at ($h = \tau = 0.01$, ($t = 2$, max $-Error = 4.2158e^{-04}$), $\alpha = 1.75$) for example 2.

Problem Test 3

3. Consider the space-fractional convection–diffusion type of equation with variable coefficients:

$$\frac{\partial u(x,t)}{\partial t} + c(x)\frac{\partial u(x,t)}{\partial x} = d(x)\frac{\partial^\alpha u(x,t)}{\partial x^\alpha} + p(x,t)$$

with initial condition

$$u(x,0) = x^2(1-x)$$

homogeneous Dirichlet boundary condition

$$u(0,t) = 0 = u(1,t)$$

with variable convection–diffusion coefficients respectively,

$$c(x) = x^{0.6}, d(x) = \Gamma(2.8)x^{3/4}$$

and the forcing function

$$p(x,t) = 2x^2(1-x)t^{1.3}/\Gamma(2.3) + 0.3x^{1.8}e^{-t}$$

The exact solution is

$$u(x,t) = x^2(1-x)e^{-t}$$

Problem test 4

4. Consider the space-fractional convection–diffusion equation with variable coefficients:

$$\frac{\partial u(x,t)}{\partial t} + c(x)\frac{\partial u(x,t)}{\partial x} = d(x)\frac{\partial^\alpha u(x,t)}{\partial x^\alpha} + p(x,t)$$

with initial condition

$$u(x,0) = x^\alpha(1-x)$$

homogeneous Dirichlet boundary condition

$$u(0,t) = 0 = u(1,t)$$

with variable convection–diffusion coefficients respectively,

$$c(x) = x^{3/5}, d(x) = x^{3/4}$$

and the forcing function

$$p(x,t) = 2x^\alpha(1-x)t^{1.3}/\Gamma(2.3) + 0.3x^{1.8}e^{-t}$$

The exact solution is

$$u(x,t) = x^\alpha(1-x)e^{-t}$$

Problem test 4 is experimented with the grid size reduction extrapolation approach stated in [23]. We have smooth enough numerical and exact solutions by using C-N scheme in Figure 6, and Table 5 shows the maximum error with the error rate is given for space fractional convection–diffusion problem with a grid size reduction extrapolation method.

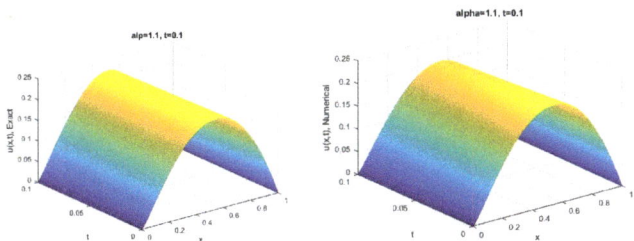

Figure 6. The exact (**left**) and numerical (**right**) solution by C-N scheme at $(h = \tau = 0.0025, (t = 0.1, \max-Error = 1.4e^{-03}, \alpha = 1.1)$ for example 4.

Table 5. The Maximum error and error-rate produced by C-N scheme for example 4 at $t = 0.1$.

		$\alpha = 1.25$		$\alpha = 1.55$	
Δt	Δx	Max-Error	Error-Rate	Max-Error	Error-Rate
1/50	1/50	1.91e−02	–	1.52e−02	–
1/100	1/100	9.9e−03	1.93	7.9e−03	1.9
1/200	1/200	5.2e−03	1.90	4.3e−03	1.84
1/400	1/400	2.8e−03	1.86	2.4e−03	1.79
1/800	1/800	1.6e−03	1.75	1.4e−03	1.7

6. Conclusions

The one dimension space fractional diffusion and fractional convection–diffusion problem with space variable coefficients is solved by the fractional C-N scheme based on the Extrapolation to the limit approach of right shifted Grünwald–Letnikov approximation. The fractional C-N method, for the fractional diffusion problem and fractional convection–diffusion equation with space variable coefficients, is consistent and unconditionally stable with second order convergence. Numerical examples confirmed that the C-N method is suitable for the space fractional convection–diffusion problem even for a large value of time domain.

Author Contributions: The authors contributed equally to the writing and approved the final manuscript of this paper. All authors have read and agreed to the published version of the manuscript.

Funding: This research was financially supported by the National Key Research (Grant No. 2017YFB0305601) and Development Program of China (Grant No. 2017YFB0701700).

Acknowledgments: The authors would like to thank the editor and the anonymous reviewers for their helpful comments for revising the article.

Conflicts of Interest: The authors declare no conflict of interest.

References

1. Diethelm, K. *The Analysis of Fractional Differential Equations*; Springer Science & Business Media: Berlin/Heidelberg, Germany, 2010.

2. Fomin, S.; Chugunov, V.; Hashida, T. Application of fractional differential equations for modeling the anomalous diffusion of contaminant from fracture into porous rock matrix with bordering alteration zone. *Trans. Porous Med.* **2010**, *81*, 187–205. [CrossRef]
3. Guo, B.; Pu, X.; Huang, F. *Fractional Partial Differential Equations and Their Numerical Solutions*; World Scientific: Singapore, 2015.
4. Podlubny, I. *Fractional Differential Equations Academic*; Academic Press: New York, NY, USA, 1999.
5. Podlubny, I. *Fractional Differential Equations: An Introduction to Fractional Derivatives, Fractional Differential Equations, to Methods of Their Solution and Some of Their Applications*; Elsevier: New York, NY, USA, 1998; Volume 198.
6. Salman, W.; Gavriilidis, A.; Angeli, P. A model for predicting axial mixing during gas–liquid Taylor flow in microchannels at low Bodenstein number. *Chem. Eng. J.* **2004**, *101*, 391–396. [CrossRef]
7. Berkowitz, B.; Cortis, A.; Dror, I.; Scher, H. Laboratory experiments on dispersive transport across interfaces. *Water Resour. Res.*, **2009**, *45*. [CrossRef]
8. Cortis, A.; Berkowitz, B. Computing "anomalous" contaminant transport in porous media. *Ground Water* **2005**, *43*, 947–950. [CrossRef]
9. Meerschaert, M.M.; Tadjeran, C. Finite difference approximations for fractional advection–dispersion flow equations. *J. Comput. Appl. Math.* **2004**, *172*, 65–77. [CrossRef]
10. Meerschaert, M.M.; Tadjeran, C. Finite difference approximations for two-sided space fractional partial differential equations. *Appl. Numer. Math.* **2006**, *56*, 80–90. [CrossRef]
11. Ren, J.; Sun, Z.Z.; Zhao, X. Compact difference scheme for the fractional sub-diffusion equation with Neumann boundary conditions. *J. Comput. Phys.* **2013**, *232*, 456–467. [CrossRef]
12. Wang, K.; Wang, H. A fast characteristic finite difference method for fractional advection–diffusion equations. *Adv. Water. Resour.* **2011**, *34*, 810–816. [CrossRef]
13. Singh, J.; Swroop, R.; Kumar, D. A computational approach for fractional convection–diffusion equation via integral transforms. *Ain Shams Eng. J.* **2016**, *9*, 1019–1028. [CrossRef]
14. Liu, F.; Zhuang, P.; Burrage, K. Numerical methods and analysis for a class of fractional advection–dispersion models. *Comput. Math. Appl.*, **2012**, *64*, 2990–3007. [CrossRef]
15. Tian, W.; Deng, W.; Wu, Y. Polynomial spectral collocation method for space fractional advection–diffusion equation. *Numer. Methods Partial Differ. Equ.* **2014**, *30*, 514–535. [CrossRef]
16. Hejazi, H.; Moroney, T.; Liu, F. Comparison of finite difference and finite volume methods for solving the space fractional advection–dispersion equation with variable coefficients. *ANZIAM J.* **2013**, *54*, 557–573. [CrossRef]
17. Liu, F.; Zhuang, P.; Turner, I.; Burrage, K.; Anh, V. A new fractional finite volume method for solving the fractional diffusion equation. *Appl. Math. Model.* **2014**, *38*, 15–16. [CrossRef]
18. Li, C.; Zeng, F. *Numerical Methods for Fractional Calculus*; Chapman and Hall/CRC: Boca Raton, FL, USA, 2015.
19. Li, C.; Zeng, F. Finite difference methods for fractional differential equations. *Int. J. Bifurcat. Chaos* **2012**, *22*, 1230014. [CrossRef]
20. Luchko, Y. Multi-dimensional fractional wave equation and some properties of its fundamental solution. *Ind. Math.* **2014**, *6*. [CrossRef]
21. Lynch, V.E.; Carreras, B.A.; del-Castillo-Negrete, D.; Ferreira-Mejias, K.M.; Hicks, H.R. Numerical methods for the solution of partial differential equations of fractional order. *J. Comput. Phys.* **2003**, *192*, 406–421. [CrossRef]
22. Sweilam, N.H.; Khader, M.M.; Mahdy, A.M.S. Crank–Nicolson finite difference method for solving time-fractional diffusion equation. *J. Fract. Calc. App.* **2012**, *2*, 1–9.
23. Tadjeran, C.; Meerschaert, M.M.; Scheffler, H.P. A second-order accurate numerical approximation for the fractional diffusion equation. *J. Comput. Phys.* **2006**, *201*, 205–213. [CrossRef]
24. Wang, Y.M. A compact finite difference method for a class of time fractional convection–diffusion-wave equations with variable coefficients. *Num. Algor.* **2015**, *70*, 625–651. [CrossRef]
25. Wang, Y.M.; Wang, T. A compact Alternative difference implicit method and its extrapolation for time fractional sub-diffusion equations with nonhomogeneous Neumann boundary conditions. *Comput. Math. Appl.* **2018**, *75*, 721–739. [CrossRef]
26. Zhou, F.; Xu, X. The third kind Chebyshev wavelets collocation method for solving the time-fractional convection–diffusion equations with variable coefficients. *Appl. Math. Comput.* **2016**, *280*, 11–29. [CrossRef]

27. Jin, B.; Lazarov, R.; Zhou, Z. A Petrov–Galerkin finite element method for fractional convection–diffusion equations. *SIAM J. Numer. Anal.* **2016**, *54*, 481–503. [CrossRef]
28. Gao, F.; Yuan, Y.; Du, N. An upwind finite volume element method for nonlinear convection–diffusion problem. *AJCM* **2011**, *1*, 264. [CrossRef]
29. Aboelenen, T. A direct discontinuous Galerkin method for fractional convection–diffusion and Schrödinger-type equations. *Eur. Phys. J. Plus* **2018**, *133*, 316. [CrossRef]
30. Xu, Q.; Hesthaven, J.S. Discontinuous Galerkin method for fractional convection–diffusion equations. *SIAM J. Numer. Anal.* **2014**, *52*, 405–423. [CrossRef]
31. Bhrawy, A.H.; Baleanu, D. A spectral Legendre–Gauss–Lobatto collocation method for a space-fractional advection–diffusion equations with variable coefficients. *Rep. Math. Phys.* **2013**, *72*, 219–233. [CrossRef]
32. Saadatmandi, A.; Dehghan, M.; Azizi, M.R. The Sinc-Legendre collocation method for a class of fractional convection–diffusion equations with variable coefficients. *Commun. Nonlinear. Sci. Numer. Simulat.* **2012**, *17*, 4125–4136. [CrossRef]
33. Pang, G.; Chen, W.; Sze, K.Y. A comparative study of finite element and finite difference methods for two-dimensional space-fractional advection–dispersion equation. *Adv. Appl. Math. Mech.* **2016**, *8*, 166–186. [CrossRef]
34. Zhang, Y. A finite difference method for fractional partial differential equation. *Appl. Math. Comput.* **2009**, *2015*, 524–529. [CrossRef]
35. Gu, X.M.; Huang, T.Z.; Ji, C.C. Carpentieri B, Alikhanov AA, Fast iterative method with a second-order implicit difference scheme for time-space fractional convection–diffusion equation. *J. Sci. Comput.* **2017**, *72*, 957–985. [CrossRef]
36. Benson, D.A.; Wheatcraft, S.; Meerschaert, M.M. Application of a fractional advection–dispersion equation. *Water Resour. Res.* **2000**, *36*, 1403–1412. [CrossRef]
37. Tian, W.; Zhou, H.; Deng, W. A class of second order difference approximations for solving space fractional diffusion equations. *Math. Comp.* **2015**, *84*, 1703–1727. [CrossRef]
38. Henrici, P. *Elements of Numerical Analysis*; John Wiley and Sons: New York, NY, USA, 1984
39. Tuan, V.K.; Gorenflo, R. Extrapolation to the Limit for Numerical Fractional Differentiation. *Z. Angew. Math. Mech.* **1995**, *75*, 646–648. [CrossRef]
40. LeVeque, R.J. *Finite Difference Methods for Ordinary and Partial Differential Equations*; SIAM: Philadelphia, PA, USA, 2007; Volume 98.
41. Isaacson, E.; Keller, H.B. *Analysis of Numerical Methods*; Courier Corporation: Chelmsford, MA, USA, 2012.

© 2020 by the authors. Licensee MDPI, Basel, Switzerland. This article is an open access article distributed under the terms and conditions of the Creative Commons Attribution (CC BY) license (http://creativecommons.org/licenses/by/4.0/).

Article

A Sharp Oscillation Criterion for a Linear Differential Equation with Variable Delay

Ábel Garab

Institute of Mathematics, University of Klagenfurt, Universitätsstraße 65–67,
9020 Klagenfurt am Wörthersee, Austria; abel.garab@aau.at

Received: 30 September 2019; Accepted: 22 October 2019; Published: 24 October 2019

Abstract: We consider linear differential equations with variable delay of the form

$$x'(t) + p(t)x(t - \tau(t)) = 0, \qquad t \geq t_0,$$

where $p \colon [t_0, \infty) \to [0, \infty)$ and $\tau \colon [t_0, \infty) \to (0, \infty)$ are continuous functions, such that $t - \tau(t) \to \infty$ (as $t \to \infty$). It is well-known that, for the oscillation of all solutions, it is necessary that

$$B := \limsup_{t \to \infty} A(t) \geq \frac{1}{e} \quad \text{holds, where} \quad A(t) := \int_{t-\tau(t)}^{t} p(s)\, ds.$$

Our main result shows that, if the function A is slowly varying at infinity (in additive form), then under mild additional assumptions on p and τ, condition $B > 1/e$ implies that all solutions of the above delay differential equation are oscillatory.

Keywords: oscillation; delay differential equation; variable delay; deviating argument; non-monotone argument; slowly varying function

MSC: 34K11; 34K06; 26A12

---

1. Introduction and Preliminary Results

Consider the following linear differential equation with variable delay:

$$x'(t) + p(t)x(t - \tau(t)) = 0, \qquad t \geq t_0, \tag{1}$$

where $p \colon [t_0, \infty) \to [0, \infty)$ and $\tau \colon [t_0, \infty) \to (0, \infty)$ are continuous functions, such that $t - \tau(t) \to \infty$ (as $t \to \infty$). Note that $t - \tau(t)$ is not assumed to be nondecreasing. Let $t_{-1} = \inf\{s - \tau(s) : s \in [t_0, \infty)\}$ and note that $t_{-1} \in (-\infty, t_0)$ holds. Then, a continuous function $x \colon [t_{-1}, \infty) \to \mathbb{R}$ is called a *solution* of Equation (1), if it is continuously differentiable on $[t_0, \infty)$ and satisfies Equation (1) there.

Such equations, and, in general, delay differential equations with either constant or variable delay arise naturally in a multitude of models from biology, physics, engineering, chemistry and economy. For an extensive introduction to the theory of delay differential equations, we refer to the books [1,2], whereas for more on their applications we recommend the reader to study [3,4].

This paper is concerned with the oscillatory behaviour of Equation (1). By convention, a solution is called *oscillatory* if it has arbitrary large zeros and is *nonoscillatory* otherwise. Results on oscillation of retarded first order equations already appeared in the works of Johann Bernoulli [5]. The first systematic study of oscillatory and nonoscillatory behaviour of Equation (1) goes back to Myshkis [6]. He showed that, in case the functions τ and p are bounded, then

$$\inf_{t\in[t_0,\infty)} \tau(t) \inf_{t\in[t_0,\infty)} p(t) > \frac{1}{e} \qquad (2)$$

implies that all solutions of Equation (1) are oscillatory, whereas condition

$$\sup_{t\in[t\_0,\infty)} \tau(t) \sup_{t\in[t\_0,\infty)} p(t) \le \frac{1}{e} \qquad (3)$$

guarantees the existence of a nonoscillatory solution.

Since then, the question of oscillation has received much attention and many results have been published providing sufficient conditions guaranteeing that all solutions are oscillatory and others that establish the existence of a nonoscillatory solution. For more details, we refer the interested reader to monographs [7–9] and to the survey papers [10,11]. Here, we only point out some results that are most relevant from our perspective.

Ladas, Lakshmikantham and Papadakis [12] proved that all solutions of Equation (1) are oscillatory, provided

$$\limsup_{t\to\infty} \int_{t-\tau(t)}^{t} p(s)\,ds > 1, \quad t-\tau(t) \text{ is nondecreasing,} \quad \text{and} \quad p(t)>0 \text{ for all } t \ge t_0. \qquad (4)$$

The following important contribution is due to Koplatadze and Chanturija [13]. For the proof, see also e.g., Theorem 2.1.1 of [9].

Theorem 1 ([13]).

(i) If

$$\liminf_{t\to\infty} \int_{t-\tau(t)}^{t} p(s)\,ds > \frac{1}{e}, \qquad (5)$$

then all solutions of Equation (1) are oscillatory.

(ii) If

$$\limsup_{t\to\infty} \int_{t-\tau(t)}^{t} p(s)\,ds < \frac{1}{e}, \qquad (6)$$

or, more generally, if

$$\int_{t-\tau(t)}^{t} p(s)\,ds \le \frac{1}{e} \quad \text{for all large } t, \qquad (7)$$

then Equation (1) has a nonoscillatory solution.

After these central results, many works have focused on filling the gap between Conditions (2) and (3), as well as between the necessary and the sufficient conditions given by Theorem 1 and Condition (4). For more on such results, see, e.g., the recent survey by Moremedi and Stavroulakis [10].

It is worth mentioning that, in case the functions τ and p are constant, then both Conditions (5) and (2) reduce to condition $\tau p > 1/e$, which is in this case not only sufficient, but—in view of Inequality (3)—also necessary for the oscillation of all solutions. Another immediate corollary of Theorem 1 is that, if $\tau(t)$ is constant $\tau > 0$, and p is τ-periodic, then $\int_{t-\tau(t)}^{t} p(s)\,ds$ is constant and Condition (7) is sharp.

Motivated by these facts, Pituk [14] recently proved that, for constant delay τ, there is a class of functions p, for which the 'almost necessary' condition $\tau \limsup_{t\to\infty} p(t) > 1/e$ is sufficient for the oscillation of all solutions of Equation (1). More precisely, he showed in Theorem 1 of [14] that, if p is slowly varying at infinity with $\liminf_{t\to\infty} p(t) > 0$, then

$$\tau \limsup_{t\to\infty} p(t) > \frac{1}{e} \qquad (8)$$

implies that all solutions of Equation (1) are oscillatory, where a function $f\colon [t_0, \infty) \to \mathbb{R}$ is called *slowly varying at infinity* if, for every $s \geq 0$,

$$f(t+s) - f(t) \to 0 \quad \text{as } t \to \infty. \qquad (9)$$

In a subsequent paper, Pituk, Stavroulakis, and the present author [15] generalized the above result and gave a class of functions p—broader than τ-periodic—for which Condition (6) is 'almost sharp'. More precisely, the following theorem was proved.

Theorem 2 ([15])**.** *Let the function τ in Equation (1) be constant, and function p be nonnegative, bounded and uniformly continuous. Assume further that the function $t \mapsto \int_{t-\tau}^{t} p(s)\,ds$ is slowly varying at infinity. Then,*

$$\liminf_{t\to\infty} \int_{t-\tau}^{t} p(s)\,ds > 0 \quad \text{and} \quad \limsup_{t\to\infty} \int_{t-\tau}^{t} p(s)\,ds > \frac{1}{e} \qquad (10)$$

imply that all solutions of Equation (1) are oscillatory.

The purpose of this paper is to show that Theorem 2 remains valid in case of variable delay, provided τ is uniformly continuous and bounded. The proof is similar to that of Theorem 2; nevertheless, some technical difficulties also arise due to the variable delay.

In the next section, we present our main theorems and give some hints to support applicability of the results. Then, in Section 3, we provide an illustrative example. Section 4 is devoted to conclusions.

2. Results

The following theorem is our main result.

Theorem 3. *For some positive numbers M and κ, let $p\colon [t_0, \infty) \to [0, M]$ and $\tau\colon [t_0, \infty) \to (0, \kappa]$ be uniformly continuous functions, and suppose that the function*

$$A\colon [t_0 + \kappa, \infty) \to [0, \infty), \qquad A(t) := \int_{t-\tau(t)}^{t} p(s)\,ds \qquad (11)$$

is slowly varying at infinity. Then,

$$\liminf_{t\to\infty} A(t) > 0 \quad \text{and} \quad \limsup_{t\to\infty} A(t) > \frac{1}{e} \qquad (12)$$

imply that all solutions of Equation (1) are oscillatory.

Before we prove the theorem, we make some comments, mainly to support applicability of the result.

From Theorem 1, it is apparent that condition $\limsup_{t\to\infty} A(t) \geq 1/e$ is necessary for the oscillation of all solutions, so Theorem 3 is sharp in this sense. Example 9 of [15] showed that the slowly varying assumption is important: even in the constant delay case, the theorem does not hold if we omit that assumption.

We remark that uniform continuity of p and τ are guaranteed, if they are globally Lipschitz continuous, which is the case if they are differentiable with their derivatives bounded on (t_0, ∞).

Let us also devote some comments to functions that are slowly varying at infinity—we shall call them *slowly varying* for brevity.

The class of slowly varying functions was studied already by Karamata [16] in a multiplicative form. For more information about slowly varying functions and their characterization, we refer the reader to the monograph by Seneta [17]. In particular, for the relation between the two terminologies, see the remark below Theorem 1.2 in Chapter 1 of [17].

Here, let us mention only one characterization of slowly varying functions given by Pituk [14] (in the additive form, see Formula (9)): a continuous function $f \colon [t_0, \infty) \to \mathbb{R}$ is slowly varying if and only if there exists $t_1 \geq t_0$, such that f can be written in the form

$$f(t) = c(t) + d(t), \quad \text{for all } t \geq t_1, \tag{13}$$

where $c \colon [t_1, \infty) \to \mathbb{R}$ is a continuous function which tends to some finite limit as $t \to \infty$, and $d \colon [t_1, \infty) \to \mathbb{R}$ is a continuously differentiable function for which $\lim_{t \to \infty} d'(t) = 0$ holds.

The next lemma will be essential in our proof.

Lemma 1 ([13]). *Suppose that $p \colon [t_0, \infty) \to [0, \infty)$ is a continuous function satisfying*

$$\liminf_{t \to \infty} \int_{t-\tau(t)}^{t} p(s) \, ds > 0.$$

If x is an eventually positive solution of Equation (1), then, for all sufficiently large T,

$$\sup_{t \geq T} \frac{x(t - \tau(t))}{x(t)} < \infty.$$

Proof of Theorem 3. Assume to the contrary that x is an eventually positive solution and all assumptions of the theorem hold (if the solution x is eventually negative, then take the solution $-x$).

By virtue of Lemma 1, there exists $T \geq t_0 + \kappa$ such that $x(t) > 0$ holds for all $t \in T - \kappa$ and

$$K := \sup_{t \geq T} \frac{x(t - \tau(t))}{x(t)} < \infty. \tag{14}$$

Then, there exists a sequence $\{t_n\}_{n \in \mathbb{N}} \subset [T, \infty)$, such that $\lim_{n \to \infty} t_n = \infty$ and

$$\lim_{n \to \infty} A(t_n) = \limsup_{t \to \infty} A(t) =: B.$$

Let us introduce the following sequence of functions:

$$y_n(t) := \frac{x(t_n + t)}{x(t_n)}, \quad p_n(t) := p(t_n + t) \quad \text{and} \quad \tau_n(t) := \tau(t_n + t) \quad \text{for all } t \geq -\kappa \text{ and } n \in \mathbb{N}. \tag{15}$$

Then, applying (1) leads to the equation

$$y'_n(t) = \frac{x'(t_n+t)}{x(t_n)} = \frac{-p(t_n+t)x(t_n+t-\tau(t_n+t))}{x(t_n)}$$
$$= -p(t_n+t)y_n(t-\tau(t_n+t)) \qquad (16)$$
$$= -p_n(t)y_n(t-\tau_n(t)). \qquad (17)$$

Now, we would like to pass to the limit by applying the Arzelà–Ascoli theorem for the above sequences of functions $\{y_n\}_{n\in\mathbb{N}}$, $\{p_n\}_{n\in\mathbb{N}}$ and $\{\tau_n\}_{n\in\mathbb{N}}$, hence we need to establish their uniform boundedness and equicontinuity. Uniform boundedness, respectively equicontinuity of $\{p_n\}_{n\in\mathbb{N}}$ and $\{\tau_n\}_{n\in\mathbb{N}}$ follow from the boundedness, respectively uniform-continuity of functions p and τ.

It remains to check these properties for $\{y_n\}_{n\in\mathbb{N}}$. For this, note that by virtue of Equation (1) and Equation (14) we obtain that the inequality

$$x'(t_n+t) = -p(t_n+t)\frac{x(t_n+t-\tau(t_n+t))}{x(t_n+t)}x(t_n+t) \geq -KMx(t_n+t)$$

holds for all $t \geq 0$ and $n \in \mathbb{N}$. This immediately implies

$$y'_n(t) = \frac{x'(t_n+t)}{x(t_n)} \geq -\frac{KMx(t_n+t)}{x(t_n)} = -KMy_n(t).$$

As y_n is positive on $[-\kappa, \infty)$, we obtain inequalities

$$-KM \leq \frac{y'_n(t)}{y_n(t)} \leq 0 \quad \text{for all } t \geq 0 \text{ and } n \in \mathbb{N}. \qquad (18)$$

Integration leads to

$$-KMt \leq \ln\frac{y_n(t)}{y_n(0)} \leq 0 \quad \text{for all } t \geq 0 \text{ and } n \in \mathbb{N}. \qquad (19)$$

Taking into account that $y_n(0) = 1$ for all $n \in \mathbb{N}$, we obtain that

$$e^{-KMt} \leq y_n(t) \leq 1 \qquad (20)$$

holds for all $t \geq 0$ and $n \in \mathbb{N}$. Now, Inequalities (20) and (18) imply that $\{y_n\}_{n\in\mathbb{N}}$ and $\{y'_n\}_{n\in\mathbb{N}}$ are uniformly bounded on $[0, \infty)$. Furthermore, the uniform boundedness of $\{y'_n\}$ yields that functions y_n are globally Lipschitz continuous with a common Lipschitz constant, and consequently $\{y_n\}_{n\in\mathbb{N}}$ is uniformly equicontinuous.

In view of the above, by the Arzelà–Ascoli theorem, we may assume (by passing to a subsequence without changing notation) that the limits

$$y(t) := \lim_{n\to\infty} y_n(t), \quad q(t) := \lim_{n\to\infty} p_n(t) \quad \text{and} \quad \sigma(t) := \lim_{n\to\infty} \tau_n(t) \qquad (21)$$

exist and are continuous on $[0, \infty)$, and the convergence is uniform on every bounded subinterval of $[0, \infty)$. Note that

$$e^{-KMt} \leq y(t) \leq 1 \qquad (22)$$

also holds for all $t \geq 0$ and $n \in \mathbb{N}$.

Furthermore, from Equation (16), together with the uniform continuity of functions p and τ and the uniform equicontinuity of $\{y_n\}_{n\in\mathbb{N}}$, we obtain that $\{y'_n\}_{n\in\mathbb{N}}$ is also equicontinuous on $[\kappa, \infty)$. Recall that the sequence $\{y'_n\}_{n\in\mathbb{N}}$ is uniformly bounded on $[0, \infty)$. Hence, according to the Arzelà–Ascoli theorem, we may assume (after passing to a subsequence if necessary) that the limit $\lim_{n\to\infty} y'_n(t)$ exists for all $t \in [\kappa, \infty)$, and the convergence is uniform on all bounded subintervals of $[\kappa, \infty)$. This combined with the fact that $\lim_{n\to\infty} y_n(\kappa) = y(\kappa)$ yields (see, e.g., Theorem 7.17 of [18]) that

$$y'(t) = \lim_{n\to\infty} y'_n(t)$$

holds for all $t \geq \kappa$. By virtue of Equation (17),

$$y'(t) = -\lim_{n\to\infty} p_n(t) y_n(t - \tau_n(t)) \tag{23}$$

is satisfied for all $t \geq \kappa$. From Equation (21) and the (uniform) equicontinuity of $\{y_n\}_{n\in\mathbb{N}}$, one can easily derive that

$$\lim_{n\to\infty} y_n(t - \tau_n(t)) = y(t - \sigma(t))$$

holds for all $t \geq \kappa$. Thus, Inequality (22) impies that y is a positive solution of equation

$$y'(t) = -q(t) y(t - \sigma(t)). \tag{24}$$

As a final step, we will apply Theorem 1 (i) to show that every solution of Equation (24) is oscillatory, which is a contradiction. Thus, we need to verify that Equation (24) fulfils the hypotheses imposed on Equation (1) and that Inequality (5) holds.

First, observe that $q(t) \in [0, M]$ and $\sigma(t) \in [0, \kappa]$ for all $t \geq \kappa$ follow immediately from their definitions and from the assumptions on p and τ, respectively. Note that we have not yet shown that $\sigma(t)$ is positive for all t.

Next, we prove that Inequality (5) is satisfied. For this, let us fix $t \geq \kappa$ and note that, since p_n converges uniformly to q on the interval $[t - \sigma(t), t]$, we obtain

$$\int_{t-\sigma(t)}^{t} q(s)\, ds = \lim_{n\to\infty} \int_{t-\sigma(t)}^{t} p_n(s)\, ds = \lim_{n\to\infty} \left(\int_{t-\tau_n(t)}^{t} p_n(s)\, ds + \int_{t-\sigma(t)}^{t-\tau_n(t)} p_n(s)\, ds \right).$$

The functions p_n are uniformly bounded, and $\tau_n(t) \to \sigma(t)$, as $n \to \infty$, so the limit of the last integral vanishes. This in turn leads to

$$\int_{t-\sigma(t)}^{t} q(s)\, ds = \lim_{n\to\infty} \int_{t-\tau(t_n+t)}^{t} p(t_n + s)\, ds$$

$$= \lim_{n\to\infty} \int_{t_n+t-\tau(t_n+t)}^{t_n+t} p(u)\, du$$

$$= \lim_{n\to\infty} A(t_n + t) = \lim_{n\to\infty} A(t_n) = B > \frac{1}{e}.$$

Here, the last inequality and the last equality hold by assumption, whereas the last but one equality follows from the slowly varying property of A. Hence, $\int_{t-\sigma(t)}^{t} q(s)\, ds$ is constant B, and thus Inequality (5) holds.

The only condition that still needs to be verified is that σ is positive for all $t \geq \kappa$. Notice that this follows immediately from the above formulas: since

$$0 < B = \int_{t-\sigma(t)}^{t} q(s)\, ds \leq M\sigma(t)$$

holds for all $t \geq \kappa$, thus $\sigma(t) \geq B/M$ for all $t \geq \kappa$.

Therefore, Theorem 1 (i) can be applied for Equation (24) with $\tau := \sigma$, $t_0 := \kappa$ and $p := q$ to obtain that every solution of Equation (24) is oscillatory, which contradicts Inequality (22). □

The following lemma may be helpful to verify the slowly varying property of A without having to evaluate it.

Lemma 2. *For some $t_0 \in \mathbb{R}$ and positive number κ, let $p\colon [t_0, \infty) \to \mathbb{R}$ be bounded and locally integrable, and $\tau\colon [t_0, \infty) \to [-\kappa, \kappa]$ be any function. If both p and τ are slowly varying at infinity, then so is the function*

$$A\colon [t_0 + \kappa, \infty) \to \mathbb{R}, \qquad A(t) := \int_{t-\tau(t)}^{t} p(s)\, ds.$$

To prove this lemma, we first need to state the following result (see Lemma 1.1 of [17]).

Lemma 3. *If $p\colon [t_0, \infty) \to \mathbb{R}$ is Lebesgue measurable and slowly varying at infinity, then, for all finite interval I, $\sup_{s \in I} |p(t+s) - p(t)| \to 0$, as $t \to \infty$.*

Proof of Lemma 2. For $t \geq t_0 + \kappa$, we have

$$A(t) = \int_{t-\tau(t)}^{t} p(s)\, ds = \int_{-\tau(t)}^{0} p(t+u)\, du = \int_{-\tau(t)}^{0} p(t+u) - p(t)\, du + \tau(t)p(t). \tag{25}$$

From this and the triangle inequality, we obtain that, for any fixed $r \in \mathbb{R}$, the inequalities

$$\begin{aligned}
|A(t+r) - A(t)| &\leq \left| \int_{-\tau(t+r)}^{0} p(t+r+u) - p(t+r)\, du \right| + \left| \int_{-\tau(t)}^{0} p(t+u) - p(t)\, du \right| \\
&\quad + |\tau(t+r)p(t+r) - \tau(t)p(t)| \\
&\leq \int_{-\kappa}^{\kappa} |p(t+r+u) - p(t+r)|\, du + \int_{-\kappa}^{\kappa} |p(t+u) - p(t)|\, du \\
&\quad + |\tau(t+r)p(t+r) - \tau(t)p(t)| \\
&\leq 2\kappa \left(\sup_{u \in [-\kappa,\kappa]} |p(t+r+u) - p(t+r)| + \sup_{u \in [-\kappa,\kappa]} |p(t+u) - p(t)| \right) \\
&\quad + |p(t+r)||\tau(t+r) - \tau(t)| + |\tau(t)||p(t+r) - p(t)|
\end{aligned}$$

hold. Now, if we let $t \to \infty$, then the last two suprema vanish due to Lemma 3 and because p is slowly varying. On the other hand, the last two terms also tend to 0, thanks to boundedness and to the slowly varying property of functions τ and p.

Therefore, $\lim_{t \to \infty} A(t+r) - A(t) = 0$ holds for all $r \geq 0$. □

Note that, for A to be slowly varying, it is not sufficient to assume merely that at least one of p and τ is slowly varying. This is the case even under the additional assumptions of Theorem 3 on p and τ. This can be readily seen by considering examples $p \equiv 1$ and $\tau(t) = 2 + \sin t$, and $\tau \equiv \pi$ and $p(t) := 2 + \sin t$, respectively. In both cases, function A will be 2π-periodic, but nonconstant, so it cannot be slowly varying.

Our last theorem is a corollary of Lemma 2 and Theorem 3, and it gives another generalization of Theorem 1 of [14] in case p is bounded.

Theorem 4. *For some positive numbers M and κ let $p \colon [t_0, \infty) \to [0, M]$ and $\tau \colon [t_0, \infty) \to (0, \kappa]$ be continuous and slowly varying at infinity. Then Condition (12) implies that all solutions of Equation (1) are oscillatory.*

Proof. First, Lemma 2 infers that function A from Equation (11) is slowly varying. As already noted after Theorem 4 of [15], the slowly varying property together with continuity implies uniform continuity. Hence, p and τ are uniformly continuous, so Theorem 3 applies, which finishes the proof. □

Let us briefly consider the case when p is unbounded, and slowly varying. If we further assume that $p(t) > 0$ holds for large t, and τ is such that there exists some $\tau_0 \in (0, \kappa]$, for which $\liminf_{t \to \infty} \tau(t) \geq \tau_0$ holds and $t - \tau(t)$ is nondecreasing (note that Theorem 1 of [14] meets these assumptions), then, using the slowly varying property of p, it can be easily shown that $\limsup_{t \to \infty} \int_{t-\tau(t)}^{t} p(s)\, ds = \infty$. In particular, Condition (4) is fulfilled, which yields that all solutions are oscillatory regardless of Condition (12).

3. Example

Before concluding the paper, let us consider the following example, which may look a bit artificial. This is because our intention was to design it in such a way that—hopefully—no other known results could guarantee the oscillation of all solutions. Obviously, it is not possible to be aware of all the related results, and to check whether they are applicable; nevertheless, we shall exclude applicability of many classical, as well as many recent theorems.

Consider the equation

$$x'(t) + \left(\frac{1}{2\pi e} + \delta \sin \sqrt{t}\right) x\left(t - (2\pi + \varepsilon \cos \sqrt{t})\right) = 0, \qquad t \geq 0, \tag{26}$$

where $\delta \in (0, \frac{1}{2\pi e})$ and $\varepsilon \in (0, 2\pi)$ are small positive constants that will be determined later. Functions p and τ are clearly positive and bounded, so Equation (26) is a special case of Equation (1) with

$$p(t) = \frac{1}{2\pi e} + \delta \sin \sqrt{t}, \qquad \tau(t) = 2\pi + \varepsilon \cos \sqrt{t} \quad \text{and} \quad t_0 = 0.$$

Note that the functions $\sin \sqrt{t}$ and $\cos \sqrt{t}$ are slowly varying at infinity, since their derivatives vanish there (see Equation (13)). This in turn yields that both p and τ are slowly varying, and, thus, in view of Lemma 2, A is slowly varying as well.

On the other hand, a direct calculation shows that

$$A(t) = \frac{2\pi + \varepsilon \cos \sqrt{t}}{2\pi e} + \delta \int_{t-\tau(t)}^{t} \sin \sqrt{s}\, ds.$$

This immediately implies

$$\frac{2\pi - \varepsilon}{2\pi e} - \delta(2\pi + \varepsilon) \leq \liminf_{t \to \infty} A(t) \leq \limsup_{t \to \infty} A(t) \leq \frac{2\pi + \varepsilon}{2\pi e} + \delta(2\pi + \varepsilon). \tag{27}$$

Now, by setting $t_n = (2n\pi)^2$ and $t'_n = ((2n+1)\pi)^2$ for all $n \in \mathbb{N}$, we obtain that

$$A(t_n) = \frac{2\pi + \varepsilon}{2\pi e} + \delta \int_{t_n - \tau(t_n)}^{t_n} \sin \sqrt{s}\, ds \geq \frac{2\pi + \varepsilon}{2\pi e} - \delta(2\pi + \varepsilon)$$

and

$$A(t'_n) = \frac{2\pi - \varepsilon}{2\pi e} + \delta \int_{t'_n - \tau(t'_n)}^{t'_n} \sin \sqrt{s}\, ds \leq \frac{2\pi - \varepsilon}{2\pi e} + \delta(2\pi + \varepsilon)$$

hold for all $n \in \mathbb{N}$. These together with Inequalities (27) yield the estimates

$$\frac{2\pi + \varepsilon}{2\pi e} - \delta(2\pi + \varepsilon) \leq \limsup_{t \to \infty} A(t) \leq \frac{2\pi + \varepsilon}{2\pi e} + \delta(2\pi + \varepsilon)$$

and

$$\frac{2\pi - \varepsilon}{2\pi e} - \delta(2\pi + \varepsilon) \leq \liminf_{t \to \infty} A(t) \leq \frac{2\pi - \varepsilon}{2\pi e} + \delta(2\pi + \varepsilon).$$

Finally, for $\gamma > 0$, let $\varepsilon := \varepsilon(\gamma) := 4\pi e \gamma$ and $\delta := \delta(\gamma) := \frac{\gamma}{2\pi + \varepsilon}$. Then, the above estimates take the form

$$\frac{1}{e} + \gamma \leq \limsup_{t \to \infty} A(t) \leq \frac{1}{e} + 3\gamma \quad \text{and} \quad \frac{1}{e} - 3\gamma \leq \liminf_{t \to \infty} A(t) \leq \frac{1}{e} - \gamma.$$

It is now easy to see that, for all $\gamma \in (0, \frac{1}{3e})$, all assumptions of Theorem 3 (and also of Theorem 4) are fulfilled, and therefore all solutions are oscillatory. Note also that, since $\limsup_{t \to \infty} A(t) \to \frac{1}{e}$ as $\gamma \to 0^+$, and $\liminf_{t \to \infty} A(t) < \frac{1}{e}$ for all $\gamma \in (0, \frac{1}{3e})$, by choosing $\gamma > 0$ small enough we can rule out the application of Conditions (4), (5) and various other sufficient conditions for the oscillation of all solutions of Equation (26) (see e.g., conditions (C_3)–(C_{12}) from [10]). Since function τ is nonconstant, therefore neither Condition (8) nor Theorem 2 can be applied to guarantee oscillation.

4. Conclusions

It has been known for almost forty years that, for the oscillation of all solutions of equation

$$x'(t) + p(t)x(t - \tau(t)) = 0, \quad t \geq t_0,$$

it is necessary that $\limsup_{t \to \infty} A(t) \geq 1/e$ holds, where $A(t) := \int_{t-\tau(t)}^{t} p(s)\, ds$ (see [13]). In our main result (see Theorem 3), we showed that, if the function A is slowly varying at infinity (see Formula (9)), then, under mild additional assumptions on p and τ, the 'almost necessary' condition $\limsup_{t \to \infty} A(t) > 1/e$ is sufficient for the oscillation of all solutions.

In Theorem 4, we formulated a corollary of Theorem 3. The advantage of this theorem is that its assumptions can be verified more easily.

The applicability and novelty of our results were demonstrated in Section 3.

Funding: This research received no external funding.

Acknowledgments: I am grateful to Mihály Pituk for introducing this research topic to me while I was visiting him in Veszprém in the framework of the Young Scientists Mentoring Programme of the University of Klagenfurt. I sincerely thank the referees for their useful comments. I am eligible for Open Access Funding by the University of Klagenfurt.

Conflicts of Interest: The author declares no conflict of interest.

References

1. Hale, J.K.; Verduyn Lunel, S. *Introduction to Functional-Differential Equations*; Applied Mathematical Sciences; Springer: New York, NY, USA, 1993; Volume 99.
2. Diekmann, O.; van Gils, S.A.; Verduyn Lunel, S.M.; Walther, H.-O. *Delay Equations. Functional, Complex, and Nonlinear Analysis*; Applied Mathematical Sciences; Springer: New York, NY, USA, 1995; Volume 110.
3. Erneux, T. *Applied Delay Differential Equations*; Surveys and Tutorials in the Applied Mathematical Sciences; Springer: New York, NY, USA, 2009; Volume 3.
4. Smith, H. *An Introduction to Delay Differential Equations with Applications to the Life Sciences*; Texts in Applied Mathematics; Springer: New York, NY, USA, 2011; Volume 57.
5. Bernoulli, J. Meditationes de chordis vibrantibus. *Comment. Acad. Sci. Imper. Petropol* **1728**, *3*, 13–28.
6. Myshkis, A.D. Linear homogeneous differential equations of first order with deviating arguments. *Uspekhi Mat. Nauk* **1950**, *5*, 160–162. (In Russian)
7. Erbe, L.H.; Kong, Q.; Zhang, B.G. *Oscillation Theory for Functional Differential Equations*; Marcel Dekker: New York, NY, USA, 1995.
8. Győri, I.; Ladas, G. *Oscillation Theory of Delay Differential Equations with Applications*; Oxford University Press: New York, NY, USA, 1991.
9. Ladde, G.S.; Lakshmikantham, V.; Zhang, B.G. *Oscillation Theory of Differential Equations with Deviating Arguments*; Monographs and Textbooks in Pure and Applied Mathematics; Marcel Dekker, Inc.: New York, NY, USA, 1987; Volume 110.
10. Moremedi, G.M.; Stavroulakis, I.P. A survey on the oscillation of delay equations with a monotone or non-monotone argument. In *Differential in Difference Equations and Applications*; Springer Proc. Math. Stat.: Cham, Switzerland, 2018; Volume 230, pp. 441–461.
11. Sficas, Y.G.; Stavroulakis, I.P. Oscillation criteria for first-order delay equations. *Bull. Lond. Math. Soc.* **2003**, *35*, 239–246. [CrossRef]
12. Ladas, G.; Lakshmikantham, V.; Papadakis, J.S. Oscillations of higher-order retarded differential equations generated by the retarded argument. In *Delay and Functional Differential Equations and Their Applications (Proc. Conf., Park City, Utah, 1972)*; Academic Press: New York, NY, USA, 1972; pp. 219–231.
13. Koplatadze, R.G.; Chanturija, T.A. On the oscillatory and monotonic solutions of first order differential equations with deviating arguments. *Differentsial'nye Uravneniya* **1982**, *18*, 1463–1465.
14. Pituk, M. Oscillation of a linear delay differential equation with slowly varying coefficient. *Appl. Math. Lett.* **2017**, *73*, 29–36. [CrossRef]
15. Garab, Á.; Pituk, M.; Stavroulakis, I.P. A sharp oscillation criterion for linear delay differential equations. *Appl. Math. Lett.* **2019**, *93*, 58–65. [CrossRef]
16. Karamata, J. Sur un mode de croissance rigulilre des fonctions. *Mathematica (Cluj)* **1930**, *4*, 38–53.
17. Seneta, E. *Regularly Varying Functions*; Lecture Notes in Mathematics; Springer-Verlag: Berlin, Germany; New York, NY, USA, 1976; Volume 508.
18. Rudin, W. *Principles of Mathematical Analysis*, 3rd ed.; McGraw-Hill, Inc.: New York, NY, USA, 1972.

© 2019 by the author. Licensee MDPI, Basel, Switzerland. This article is an open access article distributed under the terms and conditions of the Creative Commons Attribution (CC BY) license (http://creativecommons.org/licenses/by/4.0/).

Article

Approximation of a Linear Autonomous Differential Equation with Small Delay

Áron Fehér [1,2], Lőrinc Márton [1] and Mihály Pituk [2,*]

[1] Department of Electrical Engineering, Sapientia Hungarian University of Transylvania, Corunca, 547367 Mures, Romania; fehera@ms.sapientia.ro (Á.F.); martonl@ms.sapientia.ro (L.M.)
[2] Department of Mathematics, University of Pannonia, 8200 Veszprém, Hungary
* Correspondence: pitukm@almos.uni-pannon.hu

Received: 30 September 2019; Accepted: 12 October 2019; Published: 15 October 2019

Abstract: A linear autonomous differential equation with small delay is considered in this paper. It is shown that under a smallness condition the delay differential equation is asymptotically equivalent to a linear ordinary differential equation with constant coefficients. The coefficient matrix of the ordinary differential equation is a solution of an associated matrix equation and it can be written as a limit of a sequence of matrices obtained by successive approximations. The eigenvalues of the approximating matrices converge exponentially to the dominant characteristic roots of the delay differential equation and an explicit estimate for the approximation error is given.

Keywords: delay differential equation; ordinary differential equation; asymptotic equivalence; approximation; eigenvalue

1. Introduction

Let $\mathbb{C}$ and $\mathbb{C}^{n \times n}$ denote the set of complex numbers and the n-dimensional space of complex column vectors, respectively. Given a norm $\|\cdot\|$ on $\mathbb{C}^n$, the associated induced norm on $\mathbb{C}^{n \times n}$ will be denoted by the same symbol.

We will study the linear autonomous delay differential equation

$$\dot{x}(t) = Ax(t) + Bx(t-\tau), \tag{1}$$

where $\tau > 0$, $A \in \mathbb{C}^{n \times n}$ and $B \in \mathbb{C}^{n \times n}$ is a nonzero matrix. It is well-known that if $\phi : [-\tau, 0] \to \mathbb{C}^n$ is a continuous initial function, then Equation (1) has a unique solution $x : [-\tau, \infty) \to \mathbb{C}^n$ with initial values $x(t) = \phi(t)$ for $-\tau \le t \le 0$ (see [1]). The characteristic equation of Equation (1) has the form

$$\det \Delta(\lambda) = 0, \qquad \text{where } \Delta(\lambda) = \lambda I - A - B e^{-\lambda \tau}. \tag{2}$$

Throughout the paper, we will assume that

$$\|B\| \tau e^{1+\|A\|\tau} < 1, \tag{3}$$

which may be viewed as a smallness condition on the delay τ. We will show that if (3) holds, then Equation (1) is asymptotically equivalent to the ordinary differential equation

$$\dot{x} = Mx, \tag{4}$$

where $M \in \mathbb{C}^{n \times n}$ is the unique solution of the matrix equation

$$M = A + Be^{-M\tau} \tag{5}$$

such that
$$\|M\| < \mu_0, \qquad \text{where } \mu_0 = -\tau^{-1}\ln(\|B\|\tau) > 0. \tag{6}$$

Furthermore, the coefficient matrix M in Equation (4) can be written as a limit of successive approximations
$$M = \lim_{k\to\infty} M_k, \tag{7}$$
where
$$M_0 = 0 \quad \text{and} \quad M_{k+1} = A + Be^{-M_k\tau} \quad \text{for } k = 0, 1, 2, \ldots. \tag{8}$$

The convergence in (7) is exponential and we give an estimate for the approximation error $\|M - M_k\|$. It will be shown that those characteristic roots of Equation (1) which lie in the half-plane $\operatorname{Re}\lambda > -\mu_0$ with μ_0 as in (6) coincide with the eigenvalues of matrix M. As a consequence, the above dominant characteristic roots of Equation (1) can be approximated by the eigenvalues of M_k. We give an explicit estimate for the approximation error which shows that the convergence of the eigenvalues of M_k to the dominant characteristic roots of Equation (1) is exponentially fast.

The investigation of differential equations with small delays has received much attention. Some results which are related to our study are discussed in the last section of the paper.

2. Main Results

In this section, we formulate and prove our main results which were indicated in the Introduction.

2.1. Solution of the Matrix Equation and Its Approximation

First we prove the existence and uniqueness of the solution of the matrix Equation (5) satisfying (6).

Theorem 1. *Suppose* (3) *holds. Then Equation* (5) *has a unique solution* $M \in \mathbb{C}^{n\times n}$ *such that* (6) *holds.*

Before we present the proof of Theorem 1, we establish some lemmas.

Lemma 1. *Let* $P, Q \in \mathbb{C}^{n\times n}$ *and* $\gamma = \max\{\|P\|, \|Q\|\}$. *Then*
$$\|P^k - Q^k\| \le k\gamma^{k-1}\|P - Q\| \qquad \text{for } k = 1, 2, \ldots. \tag{9}$$

Proof. We will prove by induction on k that
$$P^k - Q^k = \sum_{j=0}^{k-1} P^j(P-Q)Q^{k-1-j} \tag{10}$$
for $k = 1, 2, \ldots$. Evidently, (10) holds for $k = 1$. Suppose for induction that (10) holds for some positive integer k. Then
$$P^{k+1} - Q^{k+1} = P^k(P-Q) + (P^k - Q^k)Q$$
$$= P^k(P-Q) + \left(\sum_{j=0}^{k-1} P^j(P-Q)Q^{k-1-j}\right)Q = \sum_{j=0}^{k} P^j(P-Q)Q^{k-j}.$$

Thus, (10) holds for all k. From (10), we find that
$$\|P^k - Q^k\| \le \sum_{j=0}^{k-1} \|P\|^j\|P-Q\|\|Q\|^{k-1-j} \le \|P-Q\|\sum_{j=0}^{k-1}\gamma^j\gamma^{k-1-j} = k\gamma^{k-1}\|P-Q\|$$
for $k = 1, 2, \ldots$. □

Using Lemma 1, we can prove the following result about the distance of two matrix exponentials.

Lemma 2. *Let $P, Q \in \mathbb{C}^{n \times n}$ and $\gamma = \max\{\|P\|, \|Q\|\}$. Then*

$$\|e^P - e^Q\| \leq e^\gamma \|P - Q\|. \tag{11}$$

Proof. By the definition of the matrix exponential, we have

$$e^P - e^Q = \sum_{k=0}^{\infty} \frac{P^k}{k!} - \sum_{k=0}^{\infty} \frac{Q^k}{k!} = \sum_{k=1}^{\infty} \frac{P^k - Q^k}{k!}.$$

From this, by the application of Lemma 1, we find that

$$\|e^P - e^Q\| \leq \sum_{k=1}^{\infty} \frac{\|P^k - Q^k\|}{k!} \leq \|P - Q\| \sum_{k=1}^{\infty} \frac{k\gamma^{k-1}}{k!} = \|P - Q\| \sum_{k=1}^{\infty} \frac{\gamma^{k-1}}{(k-1)!} = e^\gamma \|P - Q\|$$

which proves (11). □

We will also need some properties of the scalar equation

$$\lambda = a + be^{\lambda\tau}. \tag{12}$$

Lemma 3. *Let $a \in [0, \infty)$, $b, \tau \in (0, \infty)$ and suppose that*

$$b\tau e^{1+a\tau} < 1. \tag{13}$$

If we let $\lambda_0 = -\tau^{-1} \ln(b\tau)$, then $\lambda_0 > 0$ and Equation (12) has a unique root $\lambda_1 \in (0, \lambda_0)$. Moreover,

$$a + be^{\lambda\tau} < \lambda \quad \text{for } \lambda \in (\lambda_1, \lambda_0] \tag{14}$$

and

$$b\tau e^{\lambda\tau} < 1 \quad \text{for } \lambda < \lambda_0. \tag{15}$$

Proof. By virtue of (13), we have $b\tau < e^{-1-a\tau} < 1$ which implies that $\ln(b\tau) < 0$ and hence $\lambda_0 > 0$. Define

$$f(\lambda) = \lambda - a - be^{\lambda\tau} \quad \text{for } \lambda \in \mathbb{R}.$$

We have

$$f'(\lambda) = 1 - b\tau e^{\lambda\tau} \quad \text{and} \quad f''(\lambda) = -b\tau^2 e^{\lambda\tau} \quad \text{for } \lambda \in \mathbb{R}.$$

It is easily seen that $f'(\lambda) = 0$ if and only if $\lambda = -\tau^{-1} \ln(b\tau) = \lambda_0$. Furthermore, (13) is equivalent to $f(\lambda_0) = -\tau^{-1} \ln(b\tau) - a - \tau^{-1} > 0$. Since $f''(\lambda) < 0$ for $\lambda \in \mathbb{R}$, f' strictly decreases on $\mathbb{R}$. In particular, $f'(\lambda) > f'(\lambda_0) = 0$ for $\lambda < \lambda_0$. Therefore, (15) holds and f strictly increases on $(-\infty, \lambda_0]$. This, together with $f(0) < 0$ and $f(\lambda_0) > 0$, implies that f and hence Equation (12) have a unique root $\lambda_1 \in (0, \lambda_0)$. Since f strictly increases on $[\lambda_1, \lambda_0]$, we have that $f(\lambda) > f(\lambda_1) = 0$ for $\lambda \in (\lambda_1, \lambda_0]$. Thus, (14) holds. □

Now we can give a proof of Theorem 1.

Proof of Theorem 1. By Lemma 3, if (3) holds, then the equation

$$\mu = \|A\| + \|B\|e^{\mu\tau} \tag{16}$$

has a unique solution $\mu_1 \in (0, \mu_0)$, where μ_0 is given by (6). Moreover,

$$\|A\| + \|B\|e^{\mu\tau} < \mu \quad \text{for } \mu \in (\mu_1, \mu_0] \tag{17}$$

and

$$\|B\|\tau e^{\mu\tau} < 1 \quad \text{for } \mu < \mu_0. \tag{18}$$

Let $\mu \in [\mu_1, \mu_0)$ be fixed. Define

$$F(M) = A + Be^{-M\tau} \quad \text{for } M \in \mathbb{C}^{n \times n} \tag{19}$$

and

$$S = \{ M \in \mathbb{C}^{n \times n} \mid \|M\| \leq \mu \}. \tag{20}$$

Clearly, S is a nonempty and closed subset of $\mathbb{C}^{n \times n}$. By virtue of (17), we have for $M \in S$,

$$\|F(M)\| \leq \|A\| + \|B\|e^{\|M\|\tau} \leq \|A\| + \|B\|e^{\mu\tau} \leq \mu. \tag{21}$$

Thus, F maps S into itself. Let $M_1, M_2 \in S$. By the application of Lemma 2, we obtain

$$\|F(M_1) - F(M_2)\| = \|B(e^{-M_1\tau} - e^{-M_2\tau})\| \leq \|B\|\|e^{-M_1\tau} - e^{-M_2\tau}\| \leq \|B\|\tau e^{\mu\tau}\|M_1 - M_2\|.$$

In view of (18), $F : S \to S$ is a contraction and hence there exists a unique $M \in S$ such that $M = F(M)$. Since $\mu \in [\mu_1, \mu_0)$ was arbitrary, this completes the proof. □

In the next theorem, we show that the unique solution of Equation (5) satisfying (6) can be written as a limit of successive approximations M_k defined by (8) and we give an estimate for the approximation error.

Theorem 2. *Suppose* (3) *holds and let* $M \in \mathbb{C}^{n \times n}$ *be the solution of Equation* (5) *satisfying* (6)*. If* $\{M_k\}_{k=0}^{\infty}$ *is the sequence of matrices defined by* (8)*, then*

$$\|M_k\| \leq \mu_1 \quad \text{for } k = 0, 1, 2, \ldots, \tag{22}$$

and

$$\|M - M_k\| \leq \mu_1 q^k \quad \text{for } k = 0, 1, 2, \ldots, \tag{23}$$

where μ_1 is the unique root of Equation (16) in the interval $(0, \mu_0)$ and $q = \|B\|\tau e^{\mu_1\tau} < 1$ (see (18)).

Proof. Note that $M_{k+1} = F(M_k)$ for $k = 0, 1, 2, \ldots$, where F is defined by Equation (19). Taking $\mu = \mu_1$ in the proof of Theorem 1, we find that $\|M\| \leq \mu_1$. Moreover, from (20) and (21), we obtain that $\|M_k\| \leq \mu_1$ for $k = 0, 1, 2, \ldots$. From this and Equations (5) and (8), by the application of Lemma 2, we obtain for $k \geq 0$,

$$\|M - M_{k+1}\| = \|B(e^{-M\tau} - e^{-M_k\tau})\| \leq \|B\|\|e^{-M\tau} - e^{-M_k\tau}\| \leq \|B\|\tau e^{\mu_1\tau}\|M - M_k\| = q\|M - M_k\|.$$

From the last inequality, it follows by easy induction on k that

$$\|M - M_k\| \leq q^k\|M - M_0\| = q^k\|M\| \leq q^k\mu_1$$

for $k = 0, 1, 2, \ldots$. □

2.2. Dominant Eigenvalues and Eigensolutions

Let us summarize some facts from the theory of linear autonomous delay differential equations (see [1,2]). By an *eigenvalue* of Equation (1), we mean an eigenvalue of the generator of the solution semigroup (see [1,2] for details). It is known that $\lambda \in \mathbb{C}$ is an eigenvalue of Equation (1) if and only if λ is a root of the characteristic equation (2). Moreover, for every $\beta \in \mathbb{R}$, Equation (1) has only finite number of eigenvalues with $\text{Re}\,\lambda > \beta$. By an *entire solution* of Equation (1), we mean a differentiable function $x : (-\infty, \infty) \to \mathbb{C}^n$ satisfying Equation (1) for all $t \in (-\infty, \infty)$. To each eigenvalue λ of Equation (1), there correspond nontrivial entire solutions of the form $p(t)e^{\lambda t}$, $t \in (-\infty, \infty)$, where $p(t)$ is a $\mathbb{C}^n$-valued polynomial in t. Such solutions are sometimes called *eigensolutions* corresponding to λ.

The following theorem shows that under the smallness condition (3) the eigenvalues of Equation (1) with $\text{Re}\,\lambda > -\mu_0$ coincide with eigenvalues of matrix M from Theorem 1 and the corresponding eigensolutions satisfy the ordinary differential Equation (4).

Theorem 3. *Suppose* (3) *holds so that* $\mu_0 = -\tau^{-1} \ln(\|B\|\tau) > 0$, *and define*

$$\Lambda = \{\, \lambda \in \mathbb{C} \mid \det \Delta(\lambda) = 0,\ \text{Re}\,\lambda > -\mu_0 \,\}.$$

Let $M \in \mathbb{C}^{n \times n}$ *be the unique solution of Equation* (5) *satisfying* (6). *Then* $\Lambda = \sigma(M)$, *where* $\sigma(M)$ *denotes the set of eigenvalues of* M. *Moreover, for every* $\lambda \in \Lambda$, *Equations* (1) *and* (4) *have the same eigensolutions corresponding to* λ.

In the sequel, the eigenvalues of Equation (1) with $\text{Re}\,\lambda > -\mu_0$ will be called *dominant*.

As a preparation for the proof of Theorem 3, we establish three lemmas. First we show that if M is a solution of the matrix Equation (5), then every solution of the ordinary differential Equation (4) is an entire solution of the delay differential Equation (1).

Lemma 4. *Let* $M \in \mathbb{C}^{n \times n}$ *be a solution of Equation* (5). *Then every* $v \in \mathbb{C}^n$, $x(t) = e^{Mt}v$, $t \in (-\infty, \infty)$, *is an entire solution of Equation* (1).

Proof. Since $e^P e^Q = e^{P+Q}$ whenever P and $Q \in \mathbb{C}^{n \times n}$ commute, from Equation (5), we find that

$$\dot{x}(t) = Me^{Mt}v = (A + Be^{-M\tau})e^{Mt}v = Ae^{Mt}v + Be^{-M\tau}e^{Mt}v = Ax(t) + Be^{M(t-\tau)}v = Ax(t) + Bx(t-\tau)$$

for $t \in (-\infty, \infty)$. □

In the following lemma, we prove the uniqueness of entire solutions of the delay differential Equation (1) with an appropriate exponential growth as $t \to -\infty$.

Lemma 5. *Suppose* (3) *holds. If* x_1 *and* x_2 *are entire solutions of Equation* (1) *with* $x_1(0) = x_2(0)$ *and such that*

$$\sup_{t \leq 0} \|x_j(t)\| e^{\mu_0 t} < \infty, \qquad j = 1, 2, \tag{24}$$

with μ_0 *as in* (6), *then* $x_1 = x_2$ *identically on* $(-\infty, \infty)$.

Proof. Define

$$C = \sup_{t \leq 0} \|x_1(t) - x_2(t)\| e^{\mu_0 t}.$$

By virtue of (24), we have that $0 \leq C < \infty$. From Equation (1), we find for $t \leq 0$,

$$x_j(t) = x_j(0) - A \int_t^0 x_j(s)\,ds - B \int_t^0 x_j(s - \tau)\,ds, \qquad j = 1, 2.$$

From this, taking into account that $x_1(0) = x_2(0)$, we obtain for $t \leq 0$,

$$\|x_1(t) - x_2(t)\| \leq \|A\| \int_t^0 \|x_1(s) - x_2(s)\| \, ds + \|B\| \int_t^0 \|x_1(s-\tau) - x_2(s-\tau)\| \, ds$$

$$\leq \|A\| C \int_t^0 e^{-\mu_0 s} \, ds + \|B\| C \int_t^0 e^{-\mu_0(s-\tau)} \, ds$$

$$= C(\|A\| + \|B\|e^{\mu_0 \tau}) \int_t^0 e^{-\mu_0 s} \, ds \leq C \frac{\|A\| + \|B\|e^{\mu_0 \tau}}{\mu_0} e^{-\mu_0 t}.$$

The last inequality implies for $t \leq 0$,

$$\|x_1(t) - x_2(t)\| e^{\mu_0 t} \leq C \frac{\|A\| + \|B\|e^{\mu_0 \tau}}{\mu_0}.$$

Hence $C \leq \kappa C$, where

$$\kappa = \frac{\|A\| + \|B\|e^{\mu_0 \tau}}{\mu_0}.$$

By virtue of (17), we have that $\kappa < 1$. Hence $C = 0$ and $x_1(t) = x_2(t)$ for $t \leq 0$. The uniqueness theorem ([1] Chapter 2, Theorem 2.3) implies that $x_1(t) = x_2(t)$ for all $t \in (-\infty, \infty)$. □

Now we show that those entire solutions of Equation (1) which satisfy the growth condition

$$\sup_{t \leq 0} \|x(t)\| e^{\mu_0 t} < \infty \quad \text{with } \mu_0 \text{ as in (6)} \tag{25}$$

coincide with the solutions of the ordinary differential Equation (4).

Lemma 6. *Suppose* (3) *holds. Then, for every $v \in \mathbb{C}^n$, Equation* (1) *has exactly one entire solution x with $x(0) = v$ and satisfying* (25) *given by*

$$x(t) = e^{Mt} v \quad \text{for } t \in (-\infty, \infty), \tag{26}$$

where $M \in \mathbb{C}^{n \times n}$ is the solution of Equation (5) *with property* (6).

Proof. By Lemma 4, x defined by Equation (26) is an entire solution of Equation (1). Moreover, from Equations (6) and (26), we find for $t \leq 0$,

$$\|x(t)\| \leq e^{\|M\| |t|} \|v\| \leq e^{\mu_0 |t|} \|v\| = e^{-\mu_0 t} \|v\|.$$

Hence $\sup_{t \leq 0} \|x(t)\| e^{\mu_0 t} \leq \|v\| < \infty$. Thus, x given by Equation (26) is an entire solution of Equation (1) with $x(0) = v$ and satisfying (25). The uniqueness follows from Lemma 5. □

Now we can give a proof of Theorem 3.

Proof of Theorem 3. Suppose that $\lambda \in \Lambda$. Since $\det \Delta(\lambda) = 0$, there exists a nonzero vector $v \in \mathbb{C}^n$ such that $\Delta(\lambda) v = 0$ and hence $x(t) = e^{\lambda t} v$, $t \in (-\infty, \infty)$, is an entire solution of Equation (1). Since $\mathrm{Re}\, \lambda > -\mu_0$, we have for $t \leq 0$,

$$\|x(t)\| = |e^{\lambda t}| \|v\| = e^{t \, \mathrm{Re}\, \lambda} \|v\| \leq e^{-\mu_0 t} \|v\|,$$

which implies (25). Thus, $x(t) = e^{\lambda t} v$ is an entire solution of (1) with $x(0) = v$ and satisfying (25). By Lemma 6, we have that $e^{\lambda t} v = e^{Mt} v$ for $t \in (-\infty, \infty)$. Hence

$$\frac{e^{\lambda t} - 1}{t} v = \frac{e^{Mt} - I}{t} v \quad \text{for } t \in \mathbb{R} \setminus \{0\}.$$

Letting $t \to 0$, we obtain $\lambda v = Mv$. This proves that $\Lambda \subset \sigma(M)$.

Now suppose that $\lambda \in \sigma(M)$. Then there exists a nonzero vector $v \in \mathbb{C}^n$ such that $Mv = \lambda v$. According to Lemma 4, $x(t) = e^{Mt}v = e^{\lambda t}v$ is an entire solution of Equation (1). Hence $\Delta(\lambda)v = 0$ which implies that $\det \Delta(\lambda) = 0$. In order to prove that $\lambda \in \Lambda$, it remains to show that $\operatorname{Re} \lambda > -\mu_0$. It is well-known that $\rho(M) \leq \|M\|$, where $\rho(M) = \sup_{\lambda \in \sigma(M)} |\lambda|$ is the spectral radius of M. This, together with (6), yields
$$|\operatorname{Re} \lambda| \leq |\lambda| \leq \rho(M) \leq \|M\| < \mu_0.$$
Therefore $\operatorname{Re} \lambda > -\mu_0$ which proves that $\sigma(M) \subset \Lambda$.

Let $\lambda \in \Lambda = \sigma(M)$. By Lemma 4, every eigensolution of the ordinary differential equation (4) corresponding to λ is an eigensolution of the delay differential equation (1). Now suppose that x is an eigensolution of the delay differential equation (1) corresponding to λ. Then $x(t) = p(t)e^{\lambda t}$, where $p(t)$ is a $\mathbb{C}^n$-valued polynomial in t. If m is the order of the polynomial p, then there exists $K > 0$ such that
$$\|p(t)\| \leq K(1 + |t|^m) \quad \text{for } t \in (-\infty, \infty).$$
Since $\operatorname{Re} \lambda > -\mu_0$, we have that $\epsilon = \operatorname{Re} \lambda + \mu_0 > 0$. From this, we find for $t \leq 0$,
$$\|x(t)\| = \|p(t)\||e^{\lambda t}| = \|p(t)\|e^{t\operatorname{Re}\lambda} \leq K(1 + |t|^m)e^{t\operatorname{Re}\lambda} = K(1+|t|^m)e^{\epsilon t}e^{-\mu_0 t}.$$
Hence
$$\|x(t)\|e^{\mu_0 t} \leq K(1 + |t|^m)e^{\epsilon t} \longrightarrow 0 \quad \text{as } t \to -\infty.$$

Thus, x is an entire solution of Equation (1) satisfying the growth condition (25). By Lemma 6, x is a solution of the ordinary differential equation (4). □

2.3. Asymptotic Equivalence

The following result from the monograph by Diekmann et al. [2] gives an asymptotic description of the solutions of Equation (1) in terms of the eigensolutions.

Proposition 1. ([2] Chapter I, Theorem 5.4) *Let $x : [-\tau, \infty) \to \mathbb{C}^{n \times n}$ be a solution of Equation (1) corresponding to some continuous initial function $\phi : [-\tau, 0] \to \mathbb{C}^n$. For any $\gamma \in \mathbb{R}$ such that $\det \Delta(\lambda) = 0$ has no roots on the vertical line $\operatorname{Re} \lambda = \gamma$, we have the asymptotic expansion*
$$x(t) = \sum_{j=1}^{l} p_j(t)e^{\lambda_j t} + o(e^{\gamma t}) \quad \text{as } t \to \infty, \tag{27}$$
where $\lambda_1, \lambda_2, \ldots, \lambda_l$ are the finitely many roots of the characteristic equation (2) with real part greater than γ and $p_j(t)$ are $\mathbb{C}^n$-valued polynomials in t of order less than the multiplicity of λ_j as a zero of $\det \Delta(\lambda)$.

Now we can formulate our main result about the asymptotic equivalence of Equations (1) and (4).

Theorem 4. *Suppose that (3) holds so that $\mu_0 = -\tau^{-1} \ln(\|B\|\tau) > 0$. Let $M \in \mathbb{C}^{n \times n}$ be the solution of Equation (5) satisfying (6). Then the following statements are valid.*

(i) Every solution of the ordinary differential equation (4) is an entire solution of the delay differential equation (1).

(ii) For every solution $x : [-\tau, \infty) \to \mathbb{C}^{n \times n}$ of the delay differential equation (1) corresponding to some continuous initial function $\phi : [-\tau, 0] \to \mathbb{C}^n$, there exists a solution $\tilde{x}$ of the ordinary differential equation (4) such that
$$x(t) = \tilde{x}(t) + o(e^{-\mu_0 t}) \quad \text{as } t \to \infty. \tag{28}$$

Proof. Conclusion (i) follows from Lemma 1. We shall prove conclusion (ii) by applying Proposition 1 with $\gamma = -\mu_0$. We need to verify that Equation (2) has no root on the vertical line $\operatorname{Re} \lambda = -\mu_0$.

Suppose for contradiction that there exists $\lambda \in \mathbb{C}$ such that $\det \Delta(\lambda) = 0$ and $\operatorname{Re} \lambda = -\mu_0$. Then there exists a nonzero vector $v \in \mathbb{C}^n$ such that $\Delta(\lambda)v = 0$ and hence $\lambda v = Av + Be^{-\lambda \tau}v$. From this, we find that

$$|\lambda|\|v\| \leq \|A\|\|v\| + \|B\|\|e^{-\lambda \tau}v\| = \|A\|\|v\| + \|B\|\|e^{-\lambda \tau}\|\|v\|$$
$$= (\|A\| + \|B\|e^{-\tau \operatorname{Re} \lambda})\|v\| = (\|A\| + \|B\|e^{\mu_0 \tau})\|v\|.$$

Hence $|\lambda| \leq \|A\| + \|B\|e^{\mu_0 \tau}$, which together with (17), yields

$$\mu_0 = |\operatorname{Re} \lambda| \leq |\lambda| \leq \|A\| + \|B\|e^{\mu_0 \tau} < \mu_0,$$

a contradiction. Thus, we can apply Proposition 1 with $\gamma = -\mu_0$, which implies that the asymptotic relation (28) holds with

$$\tilde{x}(t) = \sum_{j=1}^{l} p_j(t)e^{\lambda_j t}, \tag{29}$$

where $\lambda_1, \lambda_2, \ldots, \lambda_l$ are those eigenvalues of Equation (1) which have real part greater than $-\mu_0$ and $p_j(t)$ are $\mathbb{C}^n$-valued polynomials in t. According to Theorem 3, the eigensolutions of Equation (1) corresponding to eigenvalues with real part greater than $-\mu_0$ are solutions of the ordinary differential equation (4). Hence $\tilde{x}$ given by Equation (29) is a solution of Equation (4). □

2.4. Approximation of the Dominant Eigenvalues

We will need the following result about the distance of the eigenvalues of two matrices in terms of the norm of their difference due to Bhatia, Elsner and Krause [3].

Proposition 2. *[3, Theorem 3] Let $P, Q \in \mathbb{C}^{n \times n}$ and $\gamma = \max\{\|P\|, \|Q\|\}$. Then the eigenvalues of P and Q can be enumerated as $\lambda_1, \ldots, \lambda_n$ and $\mu_1, \ldots, \mu_n$ in such a way that*

$$\max_{1 \leq j \leq n} |\lambda_j - \mu_j| \leq 4 \cdot 2^{-1/n} n^{1/n} (2\gamma)^{1-1/n} \|P - Q\|^{1/n}. \tag{30}$$

Recall that the dominant eigenvalues of Equation (1) are those roots of Equation (2) which have real part greater than $-\mu_0$. According to Theorem 3, if (3) holds, then the dominant eigenvalues of Equation (1) coincide with the eigenvalues of M, the unique solution of Equation (5) satisfying (6). By Theorem 2, M can be approximated by the sequence of matrices $\{M_k\}_{k=0}^{\infty}$ defined by (8). As a consequence, the dominant eigenvalues of the delay differential equation (1) can be approximated by the eigenvalues of M_k. The explicit estimate (23) for $\|M - M_k\|$, combined with Proposition 2, yields the following result.

Theorem 5. *Suppose (3) holds so that the dominant eigenvalues of Equation (1) coincide with the eigenvalues $\lambda_1, \ldots, \lambda_n$ of matrix M from Theorem 1 (see Theorem 3). If $\{M_k\}_{k=0}^{\infty}$ is the sequence of matrices defined by (8), then the eigenvalues $\lambda_1^{[k]}, \ldots, \lambda_n^{[k]}$ of M_k can be renumbered such that*

$$\max_{1 \leq j \leq n} |\lambda_j - \lambda_j^{[k]}| \leq 8 \cdot 4^{-1/n} n^{1/n} \mu_1 q^{k/n}, \tag{31}$$

where μ_1 and q have the meaning from Theorem 2.

Since $q < 1$, the explicit error estimate (31) in Theorem 5 shows that under the smallness condition (3) the eigenvalues of M_k converge to the dominant eigenvalues of the delay differential equation (1) at an exponential rate as $k \to \infty$.

3. Discussion

Let us briefly mention some results which are relevant to our study. For a class of linear differential equations with small delay, Ryabov [4] introduced a family of special solutions and showed that every solution is asymptotic to some special solution as $t \to \infty$. Ryabov's result was improved by Driver [5], Jarník and Kurzweil [6]. A more precise asymptotic description was given in [7]. For further related results on asymptotic integration and stability of linear differential equations with small delays, see [8] and [9]. Some improvements and a generalization to functional differential equations in Banach spaces were given by Faria and Huang [10]. Inertial and slow manifolds for differential equations with small delays were studied by Chicone [11]. Results on minimal sets of a skew-product semiflow generated by scalar differential equations with small delay can be found in the work of Alonso, Obaya and Sanz [12]. Smith and Thieme [13] showed that nonlinear autonomous differential equations with small delay generate a monotone semiflow with respect to the exponential ordering and the monotonicity has important dynamical consequences. For the effects of small delays on the stability and control, see the paper by Hale and Verduyn Lunel [14].

The results in the above listed papers show that if the delay is small, then there are similarities between the delay differential equation and an associated ordinary differential equation. The description of the associated ordinary differential equation in general requires the knowledge of certain special solutions. Since in most cases the special solutions are not known, the above results are mainly of theoretical interest. In the present paper, in the simple case of linear autonomous differential equations with small delay, we have described the coefficient matrix of the associated ordinary differential equation. Moreover, we have shown that the coefficient matrix can be approximated by a sequence of matrices defined recursively which yields an effective method for the approximation of the dominant eigenvalues.

Author Contributions: All authors contributed equally to this research and to writing the paper.

Funding: This research was funded by the Hungarian National Research, Development and Innovation Office grant no. K120186 and Széchenyi 2020 under the EFOP-3.6.1-16-2016-00015.

Conflicts of Interest: The authors declare no conflict of interest. The funders had no role in the design of the study; in the collection, analyses, or interpretation of data; in the writing of the manuscript, or in the decision to publish the results.

References

1. Hale, J.K.; Verduyn Lunel, S.M. *Introduction to Functional Differential Equations*; Springer: New York, NY, USA, 1993.
2. Diekmann, O.; van Gils, S.A.; Verduyn Lunel, S.M.; Walther, H.-O. *Delay Equations. Functional-, Complex-, and Nonlinear Analysis*; Springer: New York, NY, USA, 1995.
3. Bhatia, R.; Elsner, L.; Krause, G. Bounds for the variation of the roots of a polynomial and the eigenvalues of matrix. *Linear Algebra Appl.* **1990**, *142*, 195–209. [CrossRef]
4. Ryabov, Y.A. Certain asymptotic properties of linear systems with small time lag. *Soviet Math.* **1965**, *3*, 153–164. (In Russian)
5. Driver, R.D. Linear differential systems with small delays. *J. Differ. Equ.* **1976**, *21*, 149–167. [CrossRef]
6. Jarník, J.; Kurzweil, J. Ryabov's special solutions of functional differential equations. *Boll. Un. Mat. Ital.* **1975**, *11*, 198–218.
7. Arino, O.; Pituk, M. More on linear differential systems with small delays. *J. Differ. Equ.* **2001**, *170*, 381–407. [CrossRef]
8. Arino, O.; Gyori, I.; Pituk, M. Asymptotically diagonal delay differential systems. *J. Math. Anal. Appl.* **1996**, *204*, 701–728. [CrossRef]
9. Gyori, I.; Pituk, M. Stability criteria for linear delay differential equations. *Differ. Integral Equ.* **1997**, *10*, 841–852.
10. Faria, T.; Huang, W. Special solutions for linear functional differential equations and asymptotic behaviour. *Differ. Integral Equ.* **2005**, *18*, 337–360.

11. Chicone, C. Inertial and slow manifolds for delay equations with small delays. *J. Differ. Equ.* **2003**, *190*, 364–406. [CrossRef]
12. Alonso, A.I.; Obaya, R.; Sanz, A.M. A note on non-autonomous scalar functional differential equations with small delay. *C. R. Math.* **2005**, *340*, 155–160. [CrossRef]
13. Smith, H.L.; Thieme, H.R. Monotone semiflows in scalar non-quasi-monotone functional differential equations. *J. Math. Anal. Appl.* **1990**, *150*, 289–306. [CrossRef]
14. Hale, J.K.; Verduyn Lunel, S.M. Effects of small delays on stability and control. *Oper. Theory Adv. Appl.* **2001**, *122*, 275–301.

© 2019 by the authors. Licensee MDPI, Basel, Switzerland. This article is an open access article distributed under the terms and conditions of the Creative Commons Attribution (CC BY) license (http://creativecommons.org/licenses/by/4.0/).

Article

Around the Model of Infection Disease: The Cauchy Matrix and Its Properties

Alexander Domoshnitsky *, Irina Volinsky * and Marina Bershadsky

Department of Mathematics, Ariel University, 40700 Ariel, Israel
* Correspondence: adom@ariel.ac.il (A.D.); irinav@ariel.ac.il (I.V.);
Tel.: +972547740208 (A.D.); +972526893328 (I.V.)

Received: 27 June 2019; Accepted: 31 July 2019; Published: 6 August 2019

Abstract: In this paper the model of infection diseases by Marchuk is considered. Mathematical questions which are important in its study are discussed. Among them there are stability of stationary points, construction of the Cauchy matrices of linearized models, estimates of solutions. The novelty we propose is in a distributed feedback control which affects the antibody concentration. We use this control in the form of an integral term and come to the analysis of nonlinear integro-differential systems. New methods for the study of stability of linearized integro-differential systems describing the model of infection diseases are proposed. Explicit conditions of the exponential stability of the stationary points characterizing the state of the healthy body are obtained. The method of the paper is based on the symmetry properties of the Cauchy matrices which allow us their construction.

Keywords: integro–differential systems; Cauchy matrix; exponential stability; distributed control

1. Introduction

In this paper we consider the Marchuk model of infection diseases

$$\begin{cases} \frac{dV}{dt} = \beta V(t) - \gamma F(t) V(t) \\ \frac{dC}{dt} = \zeta(m(t)) \alpha F(t) V(t) - \mu_c (C(t) - C^*) \\ \frac{dF}{dt} = \rho C(t) - \eta \gamma F(t) V(t) - \mu_f F(t) \\ \frac{dm}{dt} = \sigma V(t) - \mu_m m(t) \end{cases} \quad (1)$$

proposed in the book [1].

Here t is time, $V(t)$ is antigen concentration rate, $C(t)$ is the plasma cell concentration rate, $F(t)$ is the antibody concentration rate, $m(t)$ is relative features of the body, $m = 0$ for the healthy body, $\zeta(m)$ takes into account the destruction of the normal functioning of the immune system, $\zeta(0) = 1$. $\alpha, \beta, \gamma, \rho, \eta, \mu_f, \mu_m, \mu_c, C^*$ are corresponding coefficients obtained as results of laboratory experiments. Let us note their biological sense of the coefficients: β—coefficient describing the antigen activity, γ —the antigen neutralizing factor, α—stimulation factor of the immune system, ρ—rate of production of antibodies by one plasma cell, μ_f—coefficient inversely proportional to the decay time of the antibodies, μ_m—coefficient inversely proportional to the organ recovery time, i.e., the coefficient μ_m characterizes the rate of regeneration of the target organ, μ_c—coefficient of reduction of plasma cells due to ageing (inversely proportional to the lifetime), σ—constant related with a particular disease, C^*—the plasma cell concentration of the healthy body. Let us describe now the structure of the model (1). The first equation presents the block of the virus dynamics. It describes the changes in the antigen concentration rate and includes the amount of the antigen in the blood. The antigen concentration decreases as a result of the interaction with the antibodies. The immune process is characterized by the antibodies, whose concentration changes with time (destruction rate) and is described by the third equation. The amount of the antibody cells decreases as a result of interaction with antigen and also as a result of the natural destruction. However, the plasma restores the antibodies and therefore the plasma state plays an important role in the immune process. Thus, the change in the concentration rate of the plasma cell is included in several differential equations describing this model. Taking into account the healthy body level of plasma cells and their natural ageing, the term $\mu_c\,(C(t) - C^*)$ is included in the second equation of system (1). The second and third equations present the immune response dynamics. Concerning the last equation of system (1), the following can be noted: (1) the value of m increases with the antigen's concentration rate $V(t)$; (2) the maximum value of m is one, in the case of 100% organ damage or zero for a fully healthy organ.

This model was studied in many works, note, for example, the recent papers [2–6] and the bibliography therein. The adding control to stabilize the system in the neighborhood of a stationary point was proposed, for example, in [5–8]. In the works [4,9,10], the basic mathematical model that takes into account the concentrated control of the immune response is proposed.

Let us discuss a motivation and novelty of our approach. In constructing every model, the influences of various additional factors that have seemed to be nonessential were neglected. The influence effect of choosing nonlinear terms by their linearization in neighborhood of stationary solution is also neglected. Even in the frame of linearized model, only approximate values of coefficients instead of exact ones are used. Changes of these coefficients with respect to time are not usually taken into account. It looks important to estimate an influence of all these factors.

In order to make this we have to obtain estimates of the elements of the Cauchy matrix of corresponding linearized (in a neighborhood of a stationary point) system. Consider the system

$$x'(t) = P(t)x(t) + G(t),$$

where $P(t)$ is a $(n \times n)$-matrix, $G(t)$ is an n-vector. Its general solution $x(t) = col\{x_1(t), ... x_n(t)\}$ can be represented in the form (see, for example, [11])

$$x(t) = \int_0^t C(t,s)G(s)ds + C(t,0)x(0),$$

where $n \times n$-matrix $C(t,s)$ is called the Cauchy matrix. Its j-th column ($j = 1, ..., n$) for every fixed s as a function of t, is a solution of the corresponding homogeneous system

$$x'(t) = P(t)x(t),$$

satisfying the initial conditions $x_i(s) = \delta_{ij}$, where

$$\delta_{ij} = \begin{cases} 1, & i = j, \\ 0, & i \neq j, \end{cases} \quad i = 1, ..., n,$$

(see, for example, [12]). This Cauchy matrix $C(t,s)$ satisfies the following symmetric properties $C(t,s) = X(s)X^{-1}(s)$, where $X(t)$ is a fundamental matrix, $C(t,0) = C(t,s)C(s,0)$, and in the case of constant matrix $P(t) = P$, $X(t-s) = C(t,s)$ is a fundamental matrix for every $s \geq 0$. These definition and properties allow us to construct and estimate $C(t,s)$.

It can be noted that the use of information about behaviour of a disease and the immune system for a long time (defined by distributed control, for example, in the form of an integral term) looks very natural in choosing a strategy of a possible treatment. We add a distributed control in the third equation, describing the antibody concentration rate to achieve stabilization of the process in the neighborhood of stationary solution in the form

$$u(t) = \int_0^t (F(s) - F^*)\, e^{-k(t-s)} ds. \tag{2}$$

Here F^* is the antibody concentration that we wish to achieve after the treatment. It can be noted that the influence of a correspondin average value instead of $F(t) - F^*$ at the point t looks reasonable. The kernel in (2) increases the influence of the previous moments which are closer to the current moment t. Note that this control is a reasonable one from the medical point of view. We consider a corresponding integro–differential system and construct its Cauchy matrix. This allows us to estimate the influence of all notes above factors on behavior of solutions.

Note the use of distributed control in stabilization in the papers [13,14]. The goal of this paper is to demonstrate new possibilities of distributed control in the model of infection diseases through analysis of integro-differential systems. From the medical point of view, our results could be interpreted as follows: supporting the immune system we transform infection disease to a stable state of "almost healthy" body. After getting this stable state we do not stop the use of corresponding medicine allowing to hold antibody concentration rate on the higher level than in the normal conditions of a healthy body. In all these stages it is important to estimate influence of many additional factors in order to hold the process in a corresponding zone. Going solution out of this zone can be dangerous for a patient. To give an instrument for these estimations is the main goal of this paper. We propose here a simple method of analysis and estimation based on a reduction of integro–differential systems to ones of ordinary differential equations.

Our paper consists of the following parts. In Section 2 we introduce the distributed control in the Marchuk model of infection diseases and explain how the analysis of this model of the fourth order can be reduced to the analysis of a system of ordinary differential equations of the fifth order. In Section 3 the Cauchy matrix of integro–differential system is constructed and the exponential stability of a stationary point is obtained. The case of uncertain coefficient in the control is studied in Section 4 where results on the exponential stability are proposed. The influence of changes in the right-hand side on behaviour of solutions is discussed in Section 5.

2. Modified Model of Infection Deceases

Adding the control (2) in the right-hand side of the third equation of system (1) we come to the system of four equations

$$\begin{cases} \frac{dV}{dt} = \beta V(t) - \gamma F(t) V(t) \\ \frac{dC}{dt} = \zeta(m(t)) \alpha F(t) V(t) - \mu_c (C(t) - C^*) \\ \frac{dF}{dt} = \rho C - \eta \gamma F(t) V(t) - \mu_f F(t) - b \int_0^t (F(s) - F^*) e^{-k(t-s)} ds \\ \frac{dm}{dt} = \sigma V(t) - \mu_m m(t) \end{cases} \quad (3)$$

Let us consider the following system of five equations

$$\begin{cases} \frac{dV}{dt} = \beta V(t) - \gamma F(t) V(t) \\ \frac{dC}{dt} = \zeta(m(t)) \alpha F(t) V(t) - \mu_c (C(t) - C^*) \\ \frac{dF}{dt} = \rho C - \eta \gamma F(t) V(t) - \mu_f F(t) - bu(t) \\ \frac{dm}{dt} = \sigma V(t) - \mu_m m(t) \\ \frac{du}{dt} = F(t) - F^* - ku(t) \end{cases} \quad (4)$$

Lemma 1. *The solution-vector* $col(v(t), s(t), f(t), m(t))$ *of system (3) and four first components of the solution-vector* $col(v(t), s(t), f(t), m(t), u(t))$ *of system (4) considered with the condition* $u(0) = 0$ *coincide.*

The proof of Lemma 1. follows from the formula of presentation of the general solution of the scalar linear equation $\frac{du}{dt} + ku(t) = F(t) - F^*$.

Note that a similar trick was used, for example, in papers [15,16].

Following [9] we can pass to the dimensionless case.

Substituting $V(t) = v(t) V_m$, $C(t) = s(t) C^*$, $F(t) = f(t) F^*$, $u(t) = \bar{u}(t) F^*$ into (3) we obtain.

$$\begin{cases} \frac{dv}{dt} = \beta v(t) - \gamma F^* f(t) v(t) \\ \frac{ds}{dt} = \alpha V_m \frac{F^*}{C^*} \zeta(m(t)) f(t) v(t) - \mu_c (s(t) - 1) \\ \frac{df}{dt} = \frac{\rho C^*}{F^*} s(t) - \eta \gamma V_m f(t) v(t) - \mu_f f(t) - b\bar{u}(t) \\ \frac{dm}{dt} = \sigma V_m v(t) - \mu_m m(t) \\ \frac{d\bar{u}}{dt} = f(t) - 1 - k\bar{u}(t) \end{cases} \quad (5)$$

Substituting $a_1 = \beta$, $a_2 = \gamma F^*$, $a_3 = \alpha V_m \frac{F^*}{C^*}$, $a_4 = \mu_f$, $a_5 = \mu_c$, $a_6 = \sigma V_m$, $a_7 = \mu_m$, $a_8 = \eta \gamma V_m$ into (6) we come to the system

$$\begin{cases} \frac{dv}{dt} = a_1 v(t) - a_2 f(t) v(t) \\ \frac{ds}{dt} = a_3 \zeta(m(t)) f(t) v(t) - a_5 (s(t) - 1) \\ \frac{df}{dt} = a_4 (s(t) - f(t)) - a_8 f(t) v(t) - b\bar{u}(t) \\ \frac{dm}{dt} = a_6 v(t) - a_7 m(t) \\ \frac{d\bar{u}}{dt} = f(t) - 1 - k\bar{u}(t) \end{cases} \quad (6)$$

Remark 1. *It was obtained by M. Chirkov and S. Rusakov (see their method of identification of parameters, for example in [5,9]) on the basis of the laboratory data of pneumonia, that $a_1 = 0.25$; $a_2 = 8.5000332$; $a_3 = 1.792175675 \times 10^9$; $a_4 = 1.95992344 \times 10^{-7}$; $a_5 = 0.5$; $a_6 = 10$; $a_7 = 0.4$; $a_8 = 1.7 \times 10^{-3}$.*

It is clear that $v = m = \bar{u} = 0$, $s = f = 1$ is a stationary point of system (6).

Linearizing system in a neighborhood of this stationary point, we obtain the corresponding linear system

$$\begin{cases} \frac{dv}{dt} = (a_1 - a_2) v \\ \frac{ds}{dt} = a_3 \zeta(0) v - a_5 (s - 1) \\ \frac{df}{dt} = -a_8 v - a_4 (f - 1) + a_4 (s - 1) - b\bar{u}(t) \\ \frac{dm}{dt} = a_6 v - a_7 m \\ \frac{d\bar{u}}{dt} = f - 1 - k\bar{u} \end{cases}$$

where $\zeta(0) = 1$, as it was noted above. Denoting $x_1 = v$, $x_2 = s - 1$, $x_3 = f - 1$, $x_4 = m$, $x_5 = \bar{u}$, we obtain

$$\begin{cases} x_1' = (a_1 - a_2) x_1 \\ x_2' = a_3 x_1 - a_5 x_2 \\ x_3' = -a_8 x_1 + a_4 x_2 - a_4 x_3 - b x_5 \\ x_4' = a_6 x_1 - a_7 x_4 \\ x_5' = x_3 - k x_5 \end{cases} \quad (7)$$

3. Constructing the Cauchy Matrix of the System (7)

In order to estimate the values of $x_1, \ldots, x_5$ and the speed of their tending to the stationary solutions we propose below a corresponding technique. Its basis is the Cauchy matrix.

The matrix of the coefficients of system (7) is following

$$A = \begin{pmatrix} a_1 - a_2 & 0 & 0 & 0 & 0 \\ a_3 & -a_5 & 0 & 0 & 0 \\ -a_8 & a_4 & -a_4 & 0 & -b \\ a_6 & 0 & 0 & -a_7 & 0 \\ 0 & 0 & 1 & 0 & -k \end{pmatrix} \quad (8)$$

Its eigenvalues are

$$\lambda_1 = \frac{-a_4-k+\sqrt{(a_4-k)^2-4b}}{2}, \quad \lambda_2 = \frac{-a_4-k-\sqrt{(a_4-k)^2-4b}}{2}, \quad (9)$$
$$\lambda_3 = -a_7, \quad \lambda_4 = -a_5, \quad \lambda_5 = a_1 - a_2.$$

Their negativity (negativity of the real parts in the case of complex λ_1 and λ_2) leads us to the assertion on stability of the stationary point $v = m = \bar{u} = 0, s = f = 1$ of system (6).

Theorem 1. *If $k > 0, b > 0$ and $a_i, 1 \leq i \leq 8$, are real positive and different and $a_1 < a_2$, then system (7) is exponentially stable.*

Remark 2. *All steps can be done for the integro–differential system (3) and system of ordinary differential equations (4) also directly without needing to pass to the dimensionless case (6). The linearization will lead us to a corresponding analog of the linear system of ordinary differential Equation (7) with the matrix of the coefficients B. Let us discuss the medicine sense of our result. Let F_0 be the value of antibody concentration rate of the healthy body. The case of $F_0 > \frac{\beta}{\gamma}$ is considered by G.I. Marchuk in his book. In this case the stationary point $V = 0, C = C^*, F = F_0, m = 0$, is stable even without control. We can try to consider the "bad" case, where $F_0 < \frac{\beta}{\gamma}$. It is clear that system (1.1) could not be stable in this case in the neighborhood of this stationary point since $\dot{V}(t)$ increases. It means that the immune system with the antibody concentration on the level of the healthy body cannot prevent increasing antigen concentration. Our control (2) in the third equation of system (1.1) cannot help us and makes this stationary point stable. We consider another stationary point $V = 0, C = C^*, F = F^*, m = 0$. Repeating the analysis of the eigenvalues of the matrix of the coefficients B, we come to the same conclusions. Let all coefficients in system (1) be positive (this is absolutely natural assumption) and $b > 0, k > 0$, then adding the control in the form (2), where $F^* > F_0 + \frac{\beta - \gamma F_0}{\gamma}$, we can achieve the exponential stability of this new stationary point of systems (3) and (4). Actually, positivity of k, b and all coefficients $a_i (i = 1,...,8)$ is preserved, to achieve the inequality $a_1 - a_2 < 0$ we have to require the noted inequality connecting F^* and F_0. One can make a conclusion that supporting for a long time the immune system, describing by antibody concentration $F(t)$ and holding it on the level F^* can be a possible way of a treatment.*

There are three possible cases:

(1) If $(a_4 - k)^2 > 4b$, then we have two different real eigenvalues λ_1 and λ_2.
(2) If $(a_4 - k)^2 = 4b$, then we have two real and multiple eigenvalues λ_1 and λ_2.
(3) If $(a_4 - k)^2 < 4b$, we have two complex eigenvalues λ_1 and λ_2.

3.1. Constructing the Cauchy Matrix in the Case 1

Using Maple, we obtain the eigenvectors of the matrix (8):

$$\vec{v}_1 = \begin{pmatrix} 0 \\ 0 \\ -\frac{2b}{a_4-k+\sqrt{(a_4-k)^2-4b}} \\ 0 \\ 1 \end{pmatrix}, \quad \vec{v}_2 = \begin{pmatrix} 0 \\ 0 \\ -\frac{2b}{a_4-k-\sqrt{(a_4-k)^2-4b}} \\ 0 \\ 1 \end{pmatrix}, \quad \vec{v}_3 = \begin{pmatrix} 0 \\ 0 \\ 0 \\ 1 \\ 0 \end{pmatrix}, \quad (10)$$

$$\vec{v}_4 = \begin{pmatrix} 0 \\ -\frac{a_4 a_5 - a_4 k - a_5^2 + a_5 k - b}{a_4} \\ -a_5 + k \\ 0 \\ 1 \end{pmatrix}, \quad \vec{v}_5 = \begin{pmatrix} -c(a_5 + a_1 - a_2) \\ -c a_3 \\ a_1 - a_2 + k \\ -\frac{(a_5 + a_1 - a_2) a_6 c}{a_1 - a_2 + a_7} \\ 1 \end{pmatrix},$$

where $c = \frac{a_1^2 - 2a_1 a_2 + a_1 a_4 + a_1 k + a_2^2 - a_2 a_4 - a_2 k + a_4 k + b}{a_1 a_8 - a_2 a_8 - a_3 a_4 + a_5 a_8}$.

Let us denote $\alpha_{31} = -\dfrac{2b}{a_4-k+\sqrt{(a_4-k)^2-4b}}, \alpha_{32} = -\dfrac{2b}{a_4-k-\sqrt{(a_4-k)^2-4b}}, \alpha_{24} = -\dfrac{a_4a_5-a_4k-a_5^2+a_5k-b}{a_4}, \alpha_{34} = -a_5+k, \alpha_{15} = -c(a_5+a_1-a_2), \alpha_{25} = -ca_3, \alpha_{35} = a_1-a_2+k, \alpha_{45} = -\dfrac{(a_5+a_1-a_2)a_6c}{a_1-a_2+a_7}$, and define the matrix

$$B = [\vec{v}_1, \vec{v}_2, \vec{v}_3, \vec{v}_4, \vec{v}_5] = \begin{pmatrix} 0 & 0 & 0 & 0 & \alpha_{15} \\ 0 & 0 & 0 & \alpha_{24} & \alpha_{25} \\ \alpha_{31} & \alpha_{32} & 0 & \alpha_{34} & \alpha_{35} \\ 0 & 0 & 1 & 0 & \alpha_{45} \\ 1 & 1 & 0 & 1 & 1 \end{pmatrix}$$

containing eigenvectors and its inverse matrix

$$B^{-1} = \begin{pmatrix} \dfrac{\alpha_{24}(\alpha_{32}-\alpha_{35})-\alpha_{25}(\alpha_{32}-\alpha_{34})}{\alpha_{15}\alpha_{24}(\alpha_{31}-\alpha_{32})} & \dfrac{\alpha_{32}-\alpha_{34}}{\alpha_{24}(\alpha_{31}-\alpha_{32})} & \dfrac{1}{\alpha_{31}-\alpha_{32}} & 0 & -\dfrac{\alpha_{32}}{\alpha_{31}-\alpha_{32}} \\ -\dfrac{\alpha_{24}(\alpha_{31}-\alpha_{35})-\alpha_{25}(\alpha_{31}-\alpha_{34})}{\alpha_{15}\alpha_{24}(\alpha_{31}-\alpha_{32})} & -\dfrac{\alpha_{31}-\alpha_{34}}{\alpha_{24}(\alpha_{31}-\alpha_{32})} & -\dfrac{1}{\alpha_{31}-\alpha_{32}} & 0 & \dfrac{\alpha_{31}}{\alpha_{31}-\alpha_{32}} \\ -\dfrac{\alpha_{45}}{\alpha_{15}} & 0 & 0 & 1 & 0 \\ -\dfrac{\alpha_{25}}{\alpha_{15}\alpha_{24}} & \dfrac{1}{\alpha_{24}} & 0 & 0 & 0 \\ \dfrac{1}{\alpha_{15}} & 0 & 0 & 0 & 0 \end{pmatrix}.$$

Let us write now the Cauchy matrix $C(t,s)$ of the the system (7). The Cauchy matrix can be written as $C(t,s) = e^{A(t-s)}$. In our case A is diagonalized: $A = BDB^{-1}$, we have $e^{A(t-s)} = Be^{D(t-s)}B^{-1}$, where the matrix D is diagonal, containing the eigenvalues of the matrix A. The columns $\vec{C}_i(t,s)_{1\leq i\leq 5}$ of the Cauchy matrix $C(t,s)$ of system (7) are the following ones:

$$\vec{C}_1(t,s) = \dfrac{\alpha_{24}(\alpha_{32}-\alpha_{35})-\alpha_{25}(\alpha_{32}-\alpha_{34})}{\alpha_{15}\alpha_{24}(\alpha_{31}-\alpha_{32})} \begin{pmatrix} 0 \\ 0 \\ -\dfrac{2b}{a_4-k+\sqrt{(a_4-k)^2-4b}} \\ 0 \\ 1 \end{pmatrix} e^{\left(\dfrac{-a_4-k+\sqrt{(a_4-k)^2-4b}}{2}\right)(t-s)} -$$

$$\dfrac{\alpha_{24}(\alpha_{31}-\alpha_{35})-\alpha_{25}(\alpha_{31}-\alpha_{34})}{\alpha_{15}\alpha_{24}(\alpha_{31}-\alpha_{32})} \begin{pmatrix} 0 \\ 0 \\ -\dfrac{2b}{a_4-k-\sqrt{(a_4-k)^2-4b}} \\ 0 \\ 1 \end{pmatrix} e^{\left(\dfrac{-a_4-k-\sqrt{(a_4-k)^2-4b}}{2}\right)(t-s)} -$$

$$\dfrac{\alpha_{45}}{\alpha_{15}} \begin{pmatrix} 0 \\ 0 \\ 0 \\ 1 \\ 0 \end{pmatrix} e^{-a_7(t-s)} - \dfrac{\alpha_{25}}{\alpha_{15}\alpha_{24}} \begin{pmatrix} 0 \\ -\dfrac{a_4a_5-a_4k-a_5^2+a_5k-b}{a_4} \\ -a_5+k \\ 0 \\ 1 \end{pmatrix} e^{-a_5(t-s)} +$$

$$\dfrac{1}{\alpha_{15}} \begin{pmatrix} -c(a_5+a_1-a_2) \\ -ca_3 \\ a_1-a_2+k \\ -\dfrac{(a_5+a_1-a_2)a_6c}{a_1-a_2+a_7} \\ 1 \end{pmatrix} e^{(a_1-a_2)(t-s)}$$

$$\vec{C}_2(t,s) = \frac{\alpha_{32}-\alpha_{34}}{\alpha_{24}(\alpha_{31}-\alpha_{32})}\begin{pmatrix} 0 \\ 0 \\ -\frac{2b}{a_4-k+\sqrt{(a_4-k)^2-4b}} \\ 0 \\ 1 \end{pmatrix} e^{\left(\frac{-a_4-k+\sqrt{(a_4-k)^2-4b}}{2}\right)(t-s)} -$$

$$\frac{\alpha_{31}-\alpha_{34}}{\alpha_{24}(\alpha_{31}-\alpha_{32})}\begin{pmatrix} 0 \\ 0 \\ -\frac{2b}{a_4-k-\sqrt{(a_4-k)^2-4b}} \\ 0 \\ 1 \end{pmatrix} e^{\left(\frac{-a_4-k-\sqrt{(a_4-k)^2-4b}}{2}\right)(t-s)} +$$

$$\frac{1}{\alpha_{24}}\begin{pmatrix} 0 \\ -\frac{a_4 a_5 - a_4 k - a_5^2 + a_5 k - b}{a_4} \\ -a_5+k \\ 0 \\ 1 \end{pmatrix} e^{-a_5(t-s)}$$

$$\vec{C}_3(t,s) = \frac{1}{\alpha_{31}-\alpha_{32}}\begin{pmatrix} 0 \\ 0 \\ -\frac{2b}{a_4-k+\sqrt{(a_4-k)^2-4b}} \\ 0 \\ 1 \end{pmatrix} e^{\left(\frac{-a_4-k+\sqrt{(a_4-k)^2-4b}}{2}\right)(t-s)} -$$

$$\frac{1}{\alpha_{31}-\alpha_{32}}\begin{pmatrix} 0 \\ 0 \\ -\frac{2b}{a_4-k-\sqrt{(a_4-k)^2-4b}} \\ 0 \\ 1 \end{pmatrix} e^{\left(\frac{-a_4-k-\sqrt{(a_4-k)^2-4b}}{2}\right)(t-s)}$$

$$\vec{C}_4(t,s) = \begin{pmatrix} 0 \\ 0 \\ 0 \\ 1 \\ 0 \end{pmatrix} e^{-a_7(t-s)}$$

$$\vec{C}_5(t,s) = -\frac{\alpha_{32}}{\alpha_{31}-\alpha_{32}}\begin{pmatrix} 0 \\ 0 \\ -\frac{2b}{a_4-k+\sqrt{(a_4-k)^2-4b}} \\ 0 \\ 1 \end{pmatrix} e^{\left(\frac{-a_4-k+\sqrt{(a_4-k)^2-4b}}{2}\right)(t-s)} +$$

$$\frac{\alpha_{31}}{\alpha_{31}-\alpha_{32}}\begin{pmatrix} 0 \\ 0 \\ -\frac{2b}{a_4-k-\sqrt{(a_4-k)^2-4b}} \\ 0 \\ 1 \end{pmatrix} e^{\left(\frac{-a_4-k-\sqrt{(a_4-k)^2-4b}}{2}\right)(t-s)}$$

3.2. Constructing the Cauchy Matrix in the Case 2

We have the eigenvalues

$$\lambda_1 = \lambda_2 = -\frac{a_4 + k}{2}, \quad \lambda_3 = -a_7, \quad \lambda_4 = -a_5, \quad \lambda_5 = a_1 - a_2, \tag{11}$$

Consider the following set of vectors

$$\vec{v}_1 = \begin{pmatrix} 0 \\ 0 \\ -\frac{2b}{a_4-k} \\ 0 \\ 1 \end{pmatrix}, \quad \vec{v}_2 = \begin{pmatrix} 0 \\ 0 \\ 0 \\ 0 \\ \frac{2}{a_4-k} \end{pmatrix}, \quad \vec{v}_3 = \begin{pmatrix} 0 \\ 0 \\ 0 \\ 1 \\ 0 \end{pmatrix},$$

$$\vec{v}_4 = \begin{pmatrix} 0 \\ -\frac{a_4 a_5 - a_4 k - a_5^2 + a_5 k - b}{a_4} \\ -a_5 + k \\ 0 \\ 1 \end{pmatrix}, \quad \vec{v}_5 = \begin{pmatrix} -c(a_5 + a_1 - a_2) \\ -c a_3 \\ a_1 - a_2 + k \\ -\frac{(a_5 + a_1 - a_2)a_6 c}{a_1 - a_2 + a_7} \\ 1 \end{pmatrix}, \tag{12}$$

here $\vec{v}_1, \vec{v}_3, \vec{v}_4, \vec{v}_5$ are the eigenvectors of matrix (8) and $\vec{v}_2$ is a root vector for $\vec{v}_1$.

Let us denote $\beta_{31} = -\frac{2b}{a_4-k}, \beta_{52} = \frac{2}{a_4-k}, \beta_{24} = -\frac{a_4 a_5 - a_4 k - a_5^2 + a_5 k - b}{a_4}, \beta_{34} = -a_5 + k, \beta_{15} = -c(a_5 + a_1 - a_2), \beta_{25} = -ca_3, \beta_{35} = a_1 - a_2 + k, \beta_{45} = -\frac{(a_5+a_1-a_2)a_6 c}{a_1-a_2+a_7}$, and define the matrix

$$B = [\vec{v}_1, \vec{v}_2, \vec{v}_3, \vec{v}_4, \vec{v}_5] = \begin{pmatrix} 0 & 0 & 0 & 0 & \beta_{15} \\ 0 & 0 & 0 & \beta_{24} & \beta_{25} \\ \beta_{31} & 0 & 0 & \beta_{34} & \beta_{35} \\ 0 & 0 & 1 & 0 & \beta_{45} \\ 1 & \beta_{52} & 0 & 1 & 1 \end{pmatrix}$$

and its inverse matrix

$$B^{-1} = \begin{pmatrix} -\frac{\beta_{24}\beta_{35}-\beta_{25}\beta_{34}}{\beta_{31}\beta_{15}\beta_{24}} & -\frac{\beta_{34}}{\beta_{24}\beta_{31}} & \frac{1}{\beta_{31}} & 0 & 0 \\ -\frac{\beta_{24}(\beta_{31}-\beta_{35})-\beta_{25}(\beta_{31}-\beta_{34})}{\beta_{31}\beta_{52}\beta_{24}\beta_{15}} & -\frac{\beta_{31}-\beta_{34}}{\beta_{31}\beta_{24}\beta_{52}} & -\frac{1}{\beta_{31}\beta_{52}} & 0 & \frac{1}{\beta_{52}} \\ -\frac{\beta_{45}}{\beta_{15}} & 0 & 0 & 1 & 0 \\ -\frac{\beta_{25}}{\beta_{15}\beta_{24}} & \frac{1}{\beta_{24}} & 0 & 0 & 0 \\ \frac{1}{\beta_{15}} & 0 & 0 & 0 & 0 \end{pmatrix}.$$

Let us denote: $\vec{u}_1(t) = \vec{v}_1 e^{\lambda_1 t}$, $\vec{u}_2(t) = (\vec{v}_2 + t\vec{v}_1)e^{\lambda_1 t}$, $\vec{u}_i(t) = \vec{v}_i e^{\lambda_i t}$, $3 \leq i \leq 5$, $\vec{w}_j(t,s) = \vec{u}_j(t-s)$, $1 \leq j \leq 5$.

Let us build the Cauchy matrix $C(t,s) = \{\vec{C}_i(t,s)\}_{1 \leq i \leq 5}$, where $\vec{C}_i(t,s) = \sum_{j=1}^{5} b_{ji} \vec{w}_j(t,s)$, $1 \leq i \leq 5$.

We have to find b_{ji}, $1 \leq i, j \leq 5$ in this representation. Taking into account that $C(s,s) = I$, where I is the identity (5×5)-matrix, we can write: $\vec{C}_i(s,s) = \sum_{j=1}^{5} b_{ji} \vec{v}_j$, $1 \leq i \leq 5$.

Setting $i = 1, 2, 3, 4, 5$, we obtain

$$\vec{C}_1(s,s) = \sum_{j=1}^{5} b_{j1} \vec{v}_j = B \begin{pmatrix} b_{11} \\ b_{21} \\ b_{31} \\ b_{41} \\ b_{51} \end{pmatrix} = \begin{pmatrix} 1 \\ 0 \\ 0 \\ 0 \\ 0 \end{pmatrix} \Rightarrow \begin{pmatrix} b_{11} \\ b_{21} \\ b_{31} \\ b_{41} \\ b_{51} \end{pmatrix} = \begin{pmatrix} -\frac{\beta_{24}\beta_{35} - \beta_{25}\beta_{34}}{\beta_{31}\beta_{15}\beta_{24}} \\ -\frac{\beta_{24}(\beta_{31}-\beta_{35}) - \beta_{25}(\beta_{31}-\beta_{34})}{\beta_{31}\beta_{52}\beta_{24}\beta_{15}} \\ -\frac{\beta_{45}}{\beta_{15}} \\ -\frac{\beta_{25}}{\beta_{15}\beta_{24}} \\ \frac{1}{\beta_{15}} \end{pmatrix}$$

$$\vec{C}_2(s,s) = \sum_{j=1}^{5} b_{j2} \vec{v}_j = B \begin{pmatrix} b_{12} \\ b_{22} \\ b_{32} \\ b_{42} \\ b_{52} \end{pmatrix} = \begin{pmatrix} 0 \\ 1 \\ 0 \\ 0 \\ 0 \end{pmatrix} \Rightarrow \begin{pmatrix} b_{12} \\ b_{22} \\ b_{32} \\ b_{42} \\ b_{52} \end{pmatrix} = \begin{pmatrix} -\frac{\beta_{34}}{\beta_{24}\beta_{31}} \\ -\frac{\beta_{31}-\beta_{34}}{\beta_{31}\beta_{24}\beta_{52}} \\ 0 \\ \frac{1}{\beta_{24}} \\ 0 \end{pmatrix}$$

$$\vec{C}_3(s,s) = \sum_{j=1}^{5} b_{j3} \vec{v}_j = B \begin{pmatrix} b_{13} \\ b_{23} \\ b_{33} \\ b_{43} \\ b_{53} \end{pmatrix} = \begin{pmatrix} 0 \\ 0 \\ 1 \\ 0 \\ 0 \end{pmatrix} \Rightarrow \begin{pmatrix} b_{13} \\ b_{23} \\ b_{33} \\ b_{43} \\ b_{53} \end{pmatrix} = \begin{pmatrix} \frac{1}{\beta_{31}} \\ -\frac{1}{\beta_{31}\beta_{52}} \\ 0 \\ 0 \\ 0 \end{pmatrix}$$

$$\vec{C}_4(s,s) = \sum_{j=1}^{5} b_{j4} \vec{v}_j = B \begin{pmatrix} b_{14} \\ b_{24} \\ b_{34} \\ b_{44} \\ b_{54} \end{pmatrix} = \begin{pmatrix} 0 \\ 0 \\ 0 \\ 1 \\ 0 \end{pmatrix} \Rightarrow \begin{pmatrix} b_{14} \\ b_{24} \\ b_{34} \\ b_{44} \\ b_{54} \end{pmatrix} = \begin{pmatrix} 0 \\ 0 \\ 1 \\ 0 \\ 0 \end{pmatrix}$$

$$\vec{C}_5(s,s) = \sum_{j=1}^{5} b_{j5} \vec{v}_j = B \begin{pmatrix} b_{15} \\ b_{25} \\ b_{35} \\ b_{45} \\ b_{55} \end{pmatrix} = \begin{pmatrix} 0 \\ 0 \\ 0 \\ 0 \\ 1 \end{pmatrix} \Rightarrow \begin{pmatrix} b_{15} \\ b_{25} \\ b_{35} \\ b_{45} \\ b_{55} \end{pmatrix} = \begin{pmatrix} 0 \\ \frac{1}{\beta_{52}} \\ 0 \\ 0 \\ 0 \end{pmatrix}$$

Substituting the coefficients b_{ji}, $1 \leq i,j \leq 5$ into the equality $\vec{C}_i(t,s) = \sum_{j=1}^{5} b_{ji} \vec{w}_j(t,s)$, $1 \leq i \leq 5$ we obtain

$$\vec{C}_1(t,s) = -\frac{\beta_{24}\beta_{35} - \beta_{25}\beta_{34}}{\beta_{31}\beta_{15}\beta_{24}} \begin{pmatrix} 0 \\ 0 \\ -\frac{2b}{a_4-k} \\ 0 \\ 1 \end{pmatrix} e^{\left(-\frac{a_4+k}{2}\right)(t-s)} -$$

$$\frac{\beta_{24}(\beta_{31} - \beta_{35}) - \beta_{25}(\beta_{31} - \beta_{34})}{\beta_{31}\beta_{52}\beta_{24}\beta_{15}} \left[\begin{pmatrix} 0 \\ 0 \\ 0 \\ 0 \\ \frac{2}{a_4-k} \end{pmatrix} + (t-s) \begin{pmatrix} 0 \\ 0 \\ -\frac{2b}{a_4-k} \\ 0 \\ 1 \end{pmatrix} \right] e^{\left(-\frac{a_4+k}{2}\right)(t-s)} -$$

$$\frac{\beta_{45}}{\beta_{15}} \begin{pmatrix} 0 \\ 0 \\ 0 \\ 1 \\ 0 \end{pmatrix} e^{-a_7(t-s)} - \frac{\beta_{25}}{\beta_{15}\beta_{24}} \begin{pmatrix} 0 \\ -\frac{a_4 a_5 - a_4 k - a_5^2 + a_5 k - b}{a_4} \\ -a_5 + k \\ 0 \\ 1 \end{pmatrix} e^{-a_5(t-s)} +$$

$$\frac{1}{\beta_{15}} \begin{pmatrix} -c(a_5 + a_1 - a_2) \\ -ca_3 \\ a_1 - a_2 + k \\ -\frac{(a_5 + a_1 - a_2)a_6 c}{a_1 - a_2 + a_7} \\ 1 \end{pmatrix} e^{(a_1 - a_2)(t-s)}$$

$$\vec{C}_2(t,s) = -\frac{\beta_{34}}{\beta_{24}\beta_{31}} \begin{pmatrix} 0 \\ 0 \\ -\frac{2b}{a_4-k} \\ 0 \\ 1 \end{pmatrix} - \frac{\beta_{31} - \beta_{34}}{\beta_{31}\beta_{24}\beta_{52}} \left[\begin{pmatrix} 0 \\ 0 \\ 0 \\ 0 \\ \frac{2}{a_4-k} \end{pmatrix} + (t-s) \begin{pmatrix} 0 \\ 0 \\ -\frac{2b}{a_4-k} \\ 0 \\ 1 \end{pmatrix} \right] e^{\left(-\frac{a_4+k}{2}\right)(t-s)} +$$

$$\frac{1}{\beta_{24}} \begin{pmatrix} 0 \\ -\frac{a_4 a_5 - a_4 k - a_5^2 + a_5 k - b}{a_4} \\ -a_5 + k \\ 0 \\ 1 \end{pmatrix} e^{-a_5(t-s)}$$

$$\vec{C}_3(t,s) = \left[\frac{1}{\beta_{31}} \begin{pmatrix} 0 \\ 0 \\ -\frac{2b}{a_4-k} \\ 0 \\ 1 \end{pmatrix} - \frac{1}{\beta_{31}\beta_{52}} \left[\begin{pmatrix} 0 \\ 0 \\ 0 \\ 0 \\ \frac{2}{a_4-k} \end{pmatrix} + (t-s) \begin{pmatrix} 0 \\ 0 \\ -\frac{2b}{a_4-k} \\ 0 \\ 1 \end{pmatrix} \right] \right] e^{\left(-\frac{a_4+k}{2}\right)(t-s)}$$

$$\vec{C}_4(t,s) = \begin{pmatrix} 0 \\ 0 \\ 0 \\ 1 \\ 0 \end{pmatrix} e^{-a_7(t-s)}$$

$$\vec{C}_5(t,s) = \frac{1}{\beta_{52}} \left[\begin{pmatrix} 0 \\ 0 \\ 0 \\ 0 \\ \frac{2}{a_4-k} \end{pmatrix} + (t-s) \begin{pmatrix} 0 \\ 0 \\ -\frac{2b}{a_4-k} \\ 0 \\ 1 \end{pmatrix} \right] e^{\left(-\frac{a_4+k}{2}\right)(t-s)}$$

3.3. Constructing the Cauchy Matrix in the Case 3

We have the eigenvalues

$$\lambda_1 = \frac{-a_4 - k + i\sqrt{4b - (a_4 - k)^2}}{2}, \quad \lambda_2 = \frac{-a_4 - k - i\sqrt{4b - (a_4 - k)^2}}{2}, \quad (13)$$

$$\lambda_3 = -a_7, \quad \lambda_4 = -a_5, \quad \lambda_5 = a_1 - a_2,$$

$$\vec{v}_1 = \begin{pmatrix} 0 \\ 0 \\ -\frac{2b}{a_4-k+i\sqrt{4b-(a_4-k)^2}} \\ 0 \\ 1 \end{pmatrix}, \quad \vec{v}_2 = \begin{pmatrix} 0 \\ 0 \\ -\frac{2b}{a_4-k-i\sqrt{4b-(a_4-k)^2}} \\ 0 \\ 1 \end{pmatrix}, \quad \vec{v}_3 = \begin{pmatrix} 0 \\ 0 \\ 0 \\ 1 \\ 0 \end{pmatrix},$$

$$\vec{v}_4 = \begin{pmatrix} 0 \\ -\frac{a_4 a_5 - a_4 k - a_5^2 + a_5 k - b}{a_4} \\ k - a_5 \\ 0 \\ 1 \end{pmatrix}, \quad \vec{v}_5 = \begin{pmatrix} -c(a_5 + a_1 - a_2) \\ -ca_3 \\ a_1 - a_2 + k \\ -\frac{(a_5 + a_1 - a_2)a_6 c}{a_1 - a_2 + a_7} \\ 1 \end{pmatrix}, \quad (14)$$

We can write first two vector-solutions as follows:

$$\vec{u_1}(t) = \begin{pmatrix} 0 \\ 0 \\ -\frac{2b}{a_4-k+i\sqrt{4b-(a_4-k)^2}} \\ 0 \\ 1 \end{pmatrix} \cdot e^{\left(\frac{-a_4-k+i\sqrt{4b-(a_4-k)^2}}{2}\right)t} =$$

$$\begin{pmatrix} 0 \\ 0 \\ -\frac{2b}{a_4-k+i\sqrt{4b-(a_4-k)^2}} \\ 0 \\ 1 \end{pmatrix} \cdot e^{-\frac{a_4+k}{2}t} \cdot \left(\cos\left(\frac{\sqrt{4b-(a_4-k)^2}}{2}t\right) + i\sin\left(\frac{\sqrt{4b-(a_4-k)^2}}{2}t\right) \right)$$

$$\vec{u_2}(t) = \begin{pmatrix} 0 \\ 0 \\ -\frac{2b}{a_4-k-i\sqrt{4b-(a_4-k)^2}} \\ 0 \\ 1 \end{pmatrix} \cdot e^{\left(\frac{-a_4-k-i\sqrt{4b-(a_4-k)^2}}{2}\right)t} =$$

$$\begin{pmatrix} 0 \\ 0 \\ -\frac{2b}{a_4-k-i\sqrt{4b-(a_4-k)^2}} \\ 0 \\ 1 \end{pmatrix} \cdot e^{-\frac{a_4+k}{2}t} \cdot \left(\cos\left(\frac{\sqrt{4b-(a_4-k)^2}}{2}t\right) - i\sin\left(\frac{\sqrt{4b-(a_4-k)^2}}{2}t\right) \right)$$

Passing to real solutions:

$$\vec{w_1}(t) = \frac{\vec{u_1}(t) + \vec{u_2}(t)}{2} = \begin{pmatrix} 0 \\ 0 \\ \frac{k-a_4}{2} \\ 0 \\ 1 \end{pmatrix} \cdot e^{-\frac{a_4+k}{2}t} \cdot \cos\left(\frac{\sqrt{4b - (a_4 - k)^2}}{2}t\right) +$$

$$\begin{pmatrix} 0 \\ 0 \\ -\frac{\sqrt{4b-(a_4-k)^2}}{2} \\ 0 \\ 0 \end{pmatrix} \cdot e^{-\frac{a_4+k}{2}t} \cdot \sin\left(\frac{\sqrt{4b - (a_4 - k)^2}}{2}t\right)$$

$$\vec{w_2}(t) = \frac{\vec{u_1}(t) - \vec{u_2}(t)}{2i} = \begin{pmatrix} 0 \\ 0 \\ \frac{\sqrt{4b-(a_4-k)^2}}{2} \\ 0 \\ 0 \end{pmatrix} \cdot e^{-\frac{a_4+k}{2}t} \cdot \cos\left(\frac{\sqrt{4b - (a_4 - k)^2}}{2}t\right) +$$

$$\begin{pmatrix} 0 \\ 0 \\ \frac{a_4-k}{2} \\ 0 \\ 1 \end{pmatrix} \cdot e^{-\frac{a_4+k}{2}t} \cdot \sin\left(\frac{\sqrt{4b - (a_4 - k)^2}}{2}t\right)$$

$$\vec{w_3}(t) = \begin{pmatrix} 0 \\ 0 \\ 0 \\ 1 \\ 0 \end{pmatrix} \cdot e^{-a_7 t}$$

$$\vec{w_4}(t) = \begin{pmatrix} 0 \\ -\frac{a_4 a_5 - a_4 k - a_5^2 + a_5 k - b}{a_4} \\ k - a_5 \\ 0 \\ 1 \end{pmatrix} \cdot e^{-a_5 t}$$

$$\vec{w_5}(t) = \begin{pmatrix} -c(a_5 + a_1 - a_2) \\ -ca_3 \\ a_1 - a_2 + k \\ -\frac{(a_5 + a_1 - a_2)a_6 c}{a_1 - a_2 + a_7} \\ 1 \end{pmatrix} \cdot e^{(a_1 - a_2)t}$$

Let us construct now the Cauchy matrix $C(t,s) = \left\{\vec{C_i}(t,s)\right\}_{i=1,\ldots,5}$ of the system. Let us define $\vec{w_i}(t,s) = \vec{w_i}(t-s)$, then

$$\vec{C}_i(t,s) = b_{1i}\vec{w_1}(t,s) + b_{2i}\vec{w_2}(t,s) + b_{3i}\vec{w_3}(t,s) + b_{4i}\vec{w_4}(t,s) + b_{5i}\vec{w_5}(t,s) =$$

$$\left[b_{1i} \cdot \left\{ \begin{pmatrix} 0 \\ 0 \\ \frac{k-a_4}{2} \\ 0 \\ 1 \end{pmatrix} \cdot e^{-\frac{a_4+k}{2}(t-s)} \cdot \cos\left(\frac{\sqrt{4b-(a_4-k)^2}}{2}(t-s)\right) + \begin{pmatrix} 0 \\ 0 \\ -\frac{\sqrt{4b-(a_4-k)^2}}{2} \\ 0 \\ 0 \end{pmatrix} \cdot e^{-\frac{a_4+k}{2}(t-s)} \cdot \sin\left(\frac{\sqrt{4b-(a_4-k)^2}}{2}(t-s)\right) \right\} \right] +$$

$$\left[b_{2i} \cdot \left\{ \begin{pmatrix} 0 \\ 0 \\ \frac{\sqrt{4b-(a_4-k)^2}}{2} \\ 0 \\ 0 \end{pmatrix} \cdot e^{-\frac{a_4+k}{2}(t-s)} \cdot \cos\left(\frac{\sqrt{4b-(a_4-k)^2}}{2}(t-s)\right) + \begin{pmatrix} 0 \\ 0 \\ \frac{a_4-k}{2} \\ 0 \\ 1 \end{pmatrix} \cdot e^{-\frac{a_4+k}{2}(t-s)} \cdot \sin\left(\frac{\sqrt{4b-(a_4-k)^2}}{2}(t-s)\right) \right\} \right] +$$

$$b_{3i} \cdot \begin{pmatrix} 0 \\ 0 \\ 0 \\ 1 \\ 0 \end{pmatrix} \cdot e^{-a_7(t-s)} + b_{4i} \cdot \begin{pmatrix} 0 \\ -\frac{a_4 a_5 - a_4 k - a_5^2 + a_5 k - b}{a_4} \\ k - a_5 \\ 0 \\ 1 \end{pmatrix} \cdot e^{-a_5(t-s)} +$$

$$b_{5i} \cdot \begin{pmatrix} -c(a_5 + a_1 - a_2) \\ -c a_3 \\ a_1 - a_2 + k \\ -\frac{(a_5 + a_1 - a_2) a_6 c}{a_1 - a_2 + a_7} \\ 1 \end{pmatrix} \cdot e^{(a_1 - a_2)(t-s)}$$

We have to find $b_{1i}, b_{2i}, b_{3i}, b_{4i}, b_{5i}$ in this representation. Taking into account that $C(s,s) = I$, where I is the identity (5×5) matrix, we can write:

$$\vec{C}_i(s,s) = b_{1i} \cdot \begin{pmatrix} 0 \\ 0 \\ \frac{k-a_4}{2} \\ 0 \\ 1 \end{pmatrix} + b_{2i} \cdot \begin{pmatrix} 0 \\ 0 \\ \frac{\sqrt{4b-(a_4-k)^2}}{2} \\ 0 \\ 0 \end{pmatrix} + b_{3i} \cdot \begin{pmatrix} 0 \\ 0 \\ 0 \\ 1 \\ 0 \end{pmatrix} +$$

$$b_{4i} \cdot \begin{pmatrix} 0 \\ -\frac{a_4 a_5 - a_4 k - a_5^2 + a_5 k - b}{a_4} \\ k - a_5 \\ 0 \\ 1 \end{pmatrix} + b_{5i} \cdot \begin{pmatrix} -c(a_5 + a_1 - a_2) \\ -c a_3 \\ a_1 - a_2 + k \\ -\frac{(a_5 + a_1 - a_2) a_6 c}{a_1 - a_2 + a_7} \\ 1 \end{pmatrix}$$

Let us denote $\gamma_{32} = \frac{\sqrt{4b-(a_4-k)^2}}{2}$, $\gamma_{24} = -\frac{a_4a_5-a_4k-a_5^2+a_5k-b}{a_4}$, $\gamma_{15} = -c(a_5+a_1-a_2)$, $\gamma_{25} = -ca_3$, $\gamma_{35} = a_1 - a_2 + k$, $\gamma_{45} = -\frac{(a_5+a_1-a_2)a_6c}{a_1-a_2+a_7}$, and define the matrix

$$B = \begin{pmatrix} 0 & 0 & 0 & 0 & \gamma_{15} \\ 0 & 0 & 0 & \gamma_{24} & \gamma_{25} \\ \frac{k-a_4}{2} & \gamma_{32} & 0 & k-a_5 & \gamma_{35} \\ 0 & 0 & 1 & 0 & \gamma_{45} \\ 1 & 0 & 0 & 1 & 1 \end{pmatrix}$$

and its inverse matrix

$$B^{-1} = \begin{pmatrix} -\frac{\gamma_{24}-\gamma_{25}}{\gamma_{15}\gamma_{24}} & -\frac{1}{\gamma_{24}} & 0 & 0 & 1 \\ -\frac{1}{2}\frac{\gamma_{24}(2\gamma_{35}-a_4+k)+\gamma_{25}(a_4-2a_5+3k)}{\gamma_{32}\gamma_{15}\gamma_{24}} & \frac{1}{2}\frac{a_4-3k+2a_5}{\gamma_{24}\gamma_{32}} & \frac{1}{\gamma_{32}} & 0 & \frac{1}{2}\frac{a_4-k}{\gamma_{32}} \\ -\frac{\gamma_{45}}{\gamma_{15}} & 0 & 0 & 1 & 0 \\ -\frac{\gamma_{25}}{\gamma_{15}\gamma_{24}} & \frac{1}{\gamma_{24}} & 0 & 0 & 0 \\ \frac{1}{\gamma_{15}} & 0 & 0 & 0 & 0 \end{pmatrix}.$$

Let us build the Cauchy matrix $C(t,s) = \left\{\vec{C}_i(t,s)\right\}_{1 \leq i \leq 5}$, where $\vec{C}_i(t,s) = \sum_{j=1}^{5} b_{ji} \vec{w}_j(t,s)$, $1 \leq i \leq 5$.

We have to find b_{ji}, $1 \leq i, j \leq 5$ in this representation. Taking into account that $C(s,s) = I$, where I is the identity (5×5)-matrix, we can write: $\vec{C}_i(s,s) = \sum_{j=1}^{5} b_{ji} \vec{v}_j$, $1 \leq i \leq 5$.

Setting $i = 1, 2, 3, 4, 5$, we obtain

$$\vec{C}_1(s,s) = \sum_{j=1}^{5} b_{j1} \vec{v}_j = B \begin{pmatrix} b_{11} \\ b_{21} \\ b_{31} \\ b_{41} \\ b_{51} \end{pmatrix} = \begin{pmatrix} 1 \\ 0 \\ 0 \\ 0 \\ 0 \end{pmatrix} \Rightarrow \begin{pmatrix} b_{11} \\ b_{21} \\ b_{31} \\ b_{41} \\ b_{51} \end{pmatrix} = \begin{pmatrix} -\frac{\gamma_{24}-\gamma_{25}}{\gamma_{15}\gamma_{24}} \\ -\frac{1}{2}\frac{\gamma_{24}(2\gamma_{35}-a_4+k)+\gamma_{25}(a_4-2a_5+3k)}{\gamma_{32}\gamma_{15}\gamma_{24}} \\ -\frac{\gamma_{45}}{\gamma_{15}} \\ -\frac{\gamma_{25}}{\gamma_{15}\gamma_{24}} \\ \frac{1}{\gamma_{15}} \end{pmatrix}$$

$$\vec{C}_2(s,s) = \sum_{j=1}^{5} b_{j2} \vec{v}_j = B \begin{pmatrix} b_{12} \\ b_{22} \\ b_{32} \\ b_{42} \\ b_{52} \end{pmatrix} = \begin{pmatrix} 0 \\ 1 \\ 0 \\ 0 \\ 0 \end{pmatrix} \Rightarrow \begin{pmatrix} b_{12} \\ b_{22} \\ b_{32} \\ b_{42} \\ b_{52} \end{pmatrix} = \begin{pmatrix} -\frac{1}{\gamma_{24}} \\ \frac{1}{2}\frac{a_4-3k+2a_5}{\gamma_{24}\gamma_{32}} \\ 0 \\ \frac{1}{\gamma_{24}} \\ 0 \end{pmatrix}$$

$$\vec{C}_3(s,s) = \sum_{j=1}^{5} b_{j3} \vec{v}_j = B \begin{pmatrix} b_{13} \\ b_{23} \\ b_{33} \\ b_{43} \\ b_{53} \end{pmatrix} = \begin{pmatrix} 0 \\ 0 \\ 1 \\ 0 \\ 0 \end{pmatrix} \Rightarrow \begin{pmatrix} b_{13} \\ b_{23} \\ b_{33} \\ b_{43} \\ b_{53} \end{pmatrix} = \begin{pmatrix} 0 \\ \frac{1}{\gamma_{32}} \\ 0 \\ 0 \\ 0 \end{pmatrix}$$

$$\vec{C}_4(s,s) = \sum_{j=1}^{5} b_{j4} \vec{v}_j = B \begin{pmatrix} b_{14} \\ b_{24} \\ b_{34} \\ b_{44} \\ b_{54} \end{pmatrix} = \begin{pmatrix} 0 \\ 0 \\ 0 \\ 1 \\ 0 \end{pmatrix} \Rightarrow \begin{pmatrix} b_{14} \\ b_{24} \\ b_{34} \\ b_{44} \\ b_{54} \end{pmatrix} = \begin{pmatrix} 0 \\ 0 \\ 1 \\ 0 \\ 0 \end{pmatrix}$$

$$\vec{C}_5(s,s) = \sum_{j=1}^{5} b_{j5} \vec{v}_j = B \begin{pmatrix} b_{15} \\ b_{25} \\ b_{35} \\ b_{45} \\ b_{55} \end{pmatrix} = \begin{pmatrix} 0 \\ 0 \\ 0 \\ 0 \\ 1 \end{pmatrix} \Rightarrow \begin{pmatrix} b_{15} \\ b_{25} \\ b_{35} \\ b_{45} \\ b_{55} \end{pmatrix} = \begin{pmatrix} 1 \\ \frac{1}{2}\frac{a_4-k}{\gamma_{32}} \\ 0 \\ 0 \\ 0 \end{pmatrix}$$

Substituting the coefficients b_{ji}, $1 \leq i,j \leq 5$ into equality $\vec{C}_i(t,s) = \sum_{j=1}^{5} b_{ji} \vec{w}_j(t,s)$, $1 \leq i \leq 5$ we obtain

$$\vec{C}_1(t,s) = -\frac{\gamma_{24}-\gamma_{25}}{\gamma_{15}\gamma_{24}} \cdot \left[\begin{pmatrix} 0 \\ 0 \\ \frac{k-a_4}{2} \\ 0 \\ 1 \end{pmatrix} \cdot e^{-\frac{a_4+k}{2}(t-s)} \cdot \cos\left(\frac{\sqrt{4b-(a_4-k)^2}}{2}(t-s)\right) + \begin{pmatrix} 0 \\ 0 \\ -\frac{\sqrt{4b-(a_4-k)^2}}{2} \\ 0 \\ 0 \end{pmatrix} \cdot e^{-\frac{a_4+k}{2}(t-s)} \cdot \sin\left(\frac{\sqrt{4b-(a_4-k)^2}}{2}(t-s)\right) \right] -$$

$$\frac{1}{2}\frac{\gamma_{24}(2\gamma_{35}-a_4+k)+\gamma_{25}(a_4-2a_5+3k)}{\gamma_{32}\gamma_{15}\gamma_{24}} \cdot \left[\begin{pmatrix} 0 \\ 0 \\ \frac{\sqrt{4b-(a_4-k)^2}}{2} \\ 0 \\ 0 \end{pmatrix} \cdot e^{-\frac{a_4+k}{2}(t-s)} \cdot \cos\left(\frac{\sqrt{4b-(a_4-k)^2}}{2}(t-s)\right) + \begin{pmatrix} 0 \\ 0 \\ \frac{a_4-k}{2} \\ 0 \\ 1 \end{pmatrix} \cdot e^{-\frac{a_4+k}{2}(t-s)} \cdot \sin\left(\frac{\sqrt{4b-(a_4-k)^2}}{2}(t-s)\right) \right] -$$

$$\frac{\gamma_{45}}{\gamma_{15}} \cdot \begin{pmatrix} 0 \\ 0 \\ 0 \\ 1 \\ 0 \end{pmatrix} \cdot e^{-a_7(t-s)} - \frac{\gamma_{25}}{\gamma_{15}\gamma_{24}} \cdot \begin{pmatrix} 0 \\ -\frac{a_4a_5-a_4k-a_5^2+a_5k-b}{a_4} \\ a_5-k \\ 0 \\ 1 \end{pmatrix} \cdot e^{-a_5(t-s)} +$$

$$\frac{1}{\gamma_{15}} \cdot \begin{pmatrix} -c(a_5+a_1-a_2) \\ -ca_3 \\ a_1-a_2+k \\ -\frac{(a_5+a_1-a_2)a_6c}{a_1-a_2+a_7} \\ 1 \end{pmatrix} \cdot e^{(a_1-a_2)(t-s)}$$

$$\vec{C}_2(t,s) = -\frac{1}{\gamma_{24}} \cdot \left[\begin{pmatrix} 0 \\ 0 \\ \frac{k-a_4}{2} \\ 0 \\ 1 \end{pmatrix} \cdot e^{-\frac{a_4+k}{2}(t-s)} \cdot \cos\left(\frac{\sqrt{4b-(a_4-k)^2}}{2}(t-s)\right) + \begin{pmatrix} 0 \\ 0 \\ -\frac{\sqrt{4b-(a_4-k)^2}}{2} \\ 0 \\ 0 \end{pmatrix} \cdot e^{-\frac{a_4+k}{2}(t-s)} \cdot \sin\left(\frac{\sqrt{4b-(a_4-k)^2}}{2}(t-s)\right) \right] +$$

$$\frac{1}{2}\frac{a_4-3k+2a_5}{\gamma_{24}\gamma_{32}} \cdot \left[\begin{pmatrix} 0 \\ 0 \\ \frac{\sqrt{4b-(a_4-k)^2}}{2} \\ 0 \\ 0 \end{pmatrix} \cdot e^{-\frac{a_4+k}{2}(t-s)} \cdot \cos\left(\frac{\sqrt{4b-(a_4-k)^2}}{2}(t-s)\right) + \begin{pmatrix} 0 \\ 0 \\ \frac{a_4-k}{2} \\ 0 \\ 1 \end{pmatrix} \cdot e^{-\frac{a_4+k}{2}(t-s)} \cdot \sin\left(\frac{\sqrt{4b-(a_4-k)^2}}{2}(t-s)\right) \right] +$$

$$\frac{1}{\gamma_{24}} \cdot \begin{pmatrix} 0 \\ -\frac{a_4 a_5 - a_4 k - a_5^2 + a_5 k - b}{a_4} \\ a_5 - k \\ 0 \\ 1 \end{pmatrix} \cdot e^{-a_5(t-s)}$$

$$\vec{C}_3(t,s) = \frac{1}{\gamma_{32}} \cdot \left[\begin{pmatrix} 0 \\ 0 \\ \frac{\sqrt{4b-(a_4-k)^2}}{2} \\ 0 \\ 0 \end{pmatrix} \cdot e^{-\frac{a_4+k}{2}(t-s)} \cdot \cos\left(\frac{\sqrt{4b-(a_4-k)^2}}{2}(t-s)\right) + \begin{pmatrix} 0 \\ 0 \\ \frac{a_4-k}{2} \\ 0 \\ 1 \end{pmatrix} \cdot e^{-\frac{a_4+k}{2}(t-s)} \cdot \sin\left(\frac{\sqrt{4b-(a_4-k)^2}}{2}(t-s)\right) \right]$$

$$\vec{C}_4(t,s) = \begin{pmatrix} 0 \\ 0 \\ 0 \\ 1 \\ 0 \end{pmatrix} \cdot e^{-a_7(t-s)}$$

$$\frac{1}{2}\frac{a_4-k}{\gamma_{32}}\left[\begin{pmatrix}0\\0\\\frac{k-a_4}{2}\\0\\1\end{pmatrix}\cdot e^{-\frac{a_4+k}{2}(t-s)}\cdot\cos\left(\frac{\sqrt{4b-(a_4-k)^2}}{2}(t-s)\right)+\right.$$

$$\begin{pmatrix}0\\0\\-\frac{\sqrt{4b-(a_4-k)^2}}{2}\\0\\0\end{pmatrix}\cdot e^{-\frac{a_4+k}{2}(t-s)}\cdot\sin\left(\frac{\sqrt{4b-(a_4-k)^2}}{2}(t-s)\right)+$$

$$\begin{pmatrix}0\\0\\\frac{\sqrt{4b-(a_4-k)^2}}{2}\\0\\0\end{pmatrix}\cdot e^{-\frac{a_4+k}{2}(t-s)}\cdot\cos\left(\frac{\sqrt{4b-(a_4-k)^2}}{2}(t-s)\right)+$$

$$\left.\begin{pmatrix}0\\0\\\frac{a_4-k}{2}\\0\\1\end{pmatrix}\cdot e^{-\frac{a_4+k}{2}(t-s)}\cdot\sin\left(\frac{\sqrt{4b-(a_4-k)^2}}{2}(t-s)\right)\right]$$

4. System with Uncertain Coefficient in the Distributed Control

Consider the following system of equations

$$\begin{cases}\frac{dV}{dt}=\beta V(t)-\gamma F(t)V(t)\\\frac{dC}{dt}=\zeta(m(t))\alpha F(t)V(t)-\mu_c(C(t)-C^*)\\\frac{dF}{dt}=\rho C-\eta\gamma F(t)V(t)-\mu_f F(t)-(b+\triangle b(t))u(t)\\\frac{dm}{dt}=\sigma V(t)-\mu_m m(t)\\\frac{du}{dt}=F(t)-F^*-ku(t)\end{cases} \quad (15)$$

Appearing $\triangle b(t)$ in the third equation can be explained by the individual reaction of the human body on the drug. Of course sensitivity of different patients' reactions can be different and it can be variable in time. We assume below that $\triangle b(t)$ is essentially a bounded function.

This system can be rewritten in the form

$$\begin{cases}x_1'=(a_1-a_2)x_1+g_1(x_1(t),x_3(t))\\x_2'=a_3x_1-a_5x_2+g_2(x_1(t),x_3(t))\\x_3'=-a_8x_1+a_4x_2-a_4x_3-(b+\triangle b(t))x_5+g_3(x_1(t),x_3(t))\\x_4'=a_6x_1-a_7x_4\\x_5'=x_3-kx_5\end{cases}, \quad (16)$$

where $g_i(x_1(t),x_3(t))(t)$, $1\leq i\leq 3$ results of "mistakes" we made in the process of the linearization.

It is clear that the model described by systems (15) and (16) were obtained under the assumption that various factors $\triangle g_i(t)$ acting on the antigen, plasma cell and antibody concentrations, were neglected. In reality these factors act although they are "small". Denote the so-called right-hand sides $G_i(t)=g_i(x_1(t),x_3(t))+\triangle g_i(t)$ for $i=1,2,3$ and $G_i(t)=\triangle g_i(t)$ for $i=4,5$. Denote $F(t)=\text{col}\{G_1(t),...,G_5(t)\}$, assume that $F(t)\in L_\infty^5$.

Consider the system

$$X'=AX+\triangle B(t)X+F(t), \quad (17)$$

where

$$X(t) = \begin{pmatrix} x_1(t) \\ x_2(t) \\ x_3(t) \\ x_4(t) \\ x_5(t) \end{pmatrix}, \quad \Delta B(t) = \begin{pmatrix} 0 & 0 & 0 & 0 & 0 \\ 0 & 0 & 0 & 0 & 0 \\ 0 & 0 & 0 & 0 & -\Delta b(t) \\ 0 & 0 & 0 & 0 & 0 \\ 0 & 0 & 0 & 0 & 0 \end{pmatrix}.$$

The natural problem is to estimate an influence of the right-hand side $F(t)$ on the solution $X(t)$.
The general solution of the system

$$X' - AX = Z \tag{18}$$

can be represented in the following form (see, for example, [11,12])

$$X(t) = \int_0^t C(t,s) Z(s) \, ds + C(t,0) X(0). \tag{19}$$

Without loss of generality, $X(0) = col\{0,0,0,0,0\}$. Substituting (19) into (17) we obtain

$$Z(t) - \Delta B(t) \int_0^t C(t,s) Z(s) \, ds = F(t), \tag{20}$$

which can be written in the operator form as

$$Z(t) = (\Omega Z)(t) + F(t), \tag{21}$$

where the operator $\Omega : L_\infty^5 \to L_\infty^5$ (L_∞^5 is the space of five vector-functions with essentially bounded components) is defined by the equality

$$(\Omega Z)(t) = \Delta B(t) \int_0^t C(t,s) Z(s) \, ds.$$

Denote $||\Omega||$ the norm of the operator Ω.
Estimating $||\Omega||$ for $(a_4 - k)^2 - 4b > 0$, we obtain

$$||\Omega|| \leq \max_{1 \leq j \leq 5} \left(\operatorname{ess\,sup}_{t \geq 0} \int_0^t \sum_{i=1}^5 \left|(\Delta B(t) C(t,s))_{ij}\right| ds \right).$$

Denoting $Q_j = \operatorname{ess\,sup}_{t \geq 0} \int_0^t \sum_{i=1}^5 \left|(\Delta B(t) C(t,s))_{ij}\right| ds$ and $\Delta b^* = \operatorname{ess\,sup}_{t \geq 0} |\Delta b(t)|$, we obtain

$$Q_1 = \triangle b^* \left[\left| \frac{\alpha_{24}(\alpha_{32}-\alpha_{35})-\alpha_{25}(\alpha_{32}-\alpha_{34})}{\alpha_{15}\alpha_{24}(\alpha_{31}-\alpha_{32})} \right| \frac{1}{|\lambda_1|} + \right. $$
$$\left. \left| \frac{\alpha_{24}(\alpha_{31}-\alpha_{35})-\alpha_{25}(\alpha_{31}-\alpha_{34})}{\alpha_{15}\alpha_{24}(\alpha_{31}-\alpha_{32})} \right| \frac{1}{|\lambda_2|} + \left| \frac{\alpha_{25}}{\alpha_5\alpha_{15}\alpha_{24}} \right| \right],$$

$$Q_2 = \triangle b^* \left[\left| \frac{\alpha_{32}-\alpha_{34}}{\alpha_{24}(\alpha_{31}-\alpha_{32})} \right| \frac{1}{|\lambda_1|} + \left| \frac{\alpha_{31}-\alpha_{34}}{\alpha_{24}(\alpha_{31}-\alpha_{32})} \right| \frac{1}{|\lambda_2|} + \frac{1}{|\alpha_5\alpha_{24}|} \right],$$

$$Q_3 = \triangle b^* \left[\frac{1}{|\alpha_{31}-\alpha_{32}|} \frac{1}{|\lambda_1|} + \frac{1}{|\alpha_{31}-\alpha_{32}|} \frac{1}{|\lambda_2|} \right], \quad (22)$$

$$Q_4 = 0,$$

$$Q_5 = \triangle b^* \left[\left| \frac{\alpha_{32}}{\alpha_{31}-\alpha_{32}} \right| \frac{1}{|\lambda_1|} + \left| \frac{\alpha_{31}}{\alpha_{31}-\alpha_{32}} \right| \frac{1}{|\lambda_2|} \right].$$

Theorem 2. *Let the assumption of Theorem 1 be fulfilled, $(a_4-k)^2 > 4b$ and the inequality $\max_{1 \leq j \leq 5} \{|Q_j|\} < 1$ be true. Then system (16) is exponential stable.*

Proof. The inequality in the condition of Theorem 2 implies that the norm $\|\Omega\|$ of the operator Ω is less than one. In this case there exists the inverse operator $(I-\Omega)^{-1}: L_\infty^5 \longrightarrow L_\infty^5$ and $Z = (I-\Omega)^{-1} F = (I + \Omega + \Omega^2 + ...) F$. It is clear that $\|Z\|_{L_\infty^5} \leq \frac{1}{1-\|\Omega\|} \|F\|_{L_\infty^5}$. It means that all components of the solution-vector Z of system (21) are bounded. The Cauchy matrix of system (16) satisfies the exponential estimate i.e., there exist such positive N, M that

$$|C_{ij}(t,s)| \leq N e^{-M(t-s)}, 0 \leq s \leq t < \infty.$$

Then all components of the solution-vector $X(t)$ of system (17) are bounded, according to representation (19). The exponential stability of the homogeneous system

$$X'(t) = AX(t) + \triangle B(t) X(t)$$

follows now from Bohl-Perron theorem (see, for example, [11] p. 500, [12] p. 93). □

Example 1. *Substituting the values from Remark 1 and setting $k = 4, b = 1$ we obtain*

$$Q_1 \leq 327.0253788, \quad Q_2 \leq 0.000001437277837, \quad Q_3 \leq 1.154699764, \quad Q_4 = 0, \quad Q_5 \leq 0.5773500802.$$

The inequality $327.0253788 \cdot \triangle b^* < 1$ implies the inequality $\max_{1 \leq j \leq 5} \{|Q_j|\} < 1$.
Thus if, according to Theorem 2 $\triangle b^* < 0.003057866651$, then the system (16) is exponentially stable.

Let us estimate $\|\Omega\|$ for $(a_4-k)^2 = 4b$. Denoting $P_j = \text{ess sup}_{t \geq 0} \int_0^t \sum_{i=1}^5 |(\triangle B(t) C(t,s))_{ij}| ds$, we obtain

$$P_1 = \triangle b^* \left[\left| \frac{\beta_{24}\beta_{35} - \beta_{25}\beta_{34}}{\beta_{31}\beta_{15}\beta_{24}} \right| \frac{2}{|a_4+k|} + \left| \frac{\beta_{24}(\beta_{31}-\beta_{35}) - \beta_{25}(\beta_{31}-\beta_{34})}{\beta_{31}\beta_{52}\beta_{24}\beta_{15}} \right| \left[\frac{4}{|a_4^2-k^2|} + \frac{2}{|a_4+k|} \right] \right.$$
$$\left. + \left| \frac{\beta_{25}}{\beta_{15}\beta_{24}} \right| \frac{1}{|a_5|} + \frac{1}{|\beta_{15}|} \frac{1}{|a_1-a_2|} \right],$$

$$P_2 = \triangle b^* \left[\left| \frac{\beta_{34}}{\beta_{24}\beta_{31}} \right| \frac{2}{|a_4+k|} + \left| \frac{\beta_{31}-\beta_{34}}{\beta_{31}\beta_{24}\beta_{52}} \right| \left[\frac{4}{|a_4^2-k^2|} + \frac{2}{|a_4+k|} \right] + \frac{1}{|\beta_{24}|} \frac{1}{|a_5|} \right],$$

$$P_3 = \triangle b^* \left[\frac{1}{|\beta_{31}|} \frac{2}{|a_4+k|} + \frac{1}{|\beta_{31}\beta_{52}|} \left[\frac{4}{|a_4^2-k^2|} + \frac{2}{|a_4+k|} \right] \right],$$

$$P_4 = 0,$$

$$P_5 = \triangle b^* \frac{1}{|\beta_{52}|} \left[\frac{4}{|a_4^2-k^2|} + \frac{2}{|a_4+k|} \right].$$

(23)

Theorem 3. *Let the assumption of Theorem 1 be fulfilled, $(a_4 - k)^2 = 4b$ and the inequality $\max_{1 \leq j \leq 5} \{|P_j|\} < 1$ be true. Then system (16) is exponential stable.*

The proof of Theorem 3 repeats the proof of Theorem 2.

Example 2. *Substituting the values from Remark 1 and setting $k = 1, b = 0.249999902$, we obtain the inequalities*

$$P_1 \leq 4.735918812 \cdot 10^{13}, \quad P_2 \leq 2.047987177 \cdot 10^5, \quad P_3 \leq 9.999999608, \quad P_4 = 0, \quad P_5 \leq 2.999999216.$$

The inequality $4.735918812 \cdot 10^{13} \cdot \triangle b^* < 1$ implies the inequality $\max_{1 \leq j \leq 5} \{|P_j|\} < 1$.
Thus if $\triangle b^* < 2.111522684 \cdot 10^{-14}$, then the system (16) is exponentially stable, according to Theorem 3.

Let us estimate $\|\Omega\|$ for $(a_4 - k)^2 - 4b < 0$. Denoting $R_j = \operatorname{ess\,sup}_{t \geq 0} \int_0^t \sum_{i=1}^5 \left| (\Delta B(t) C(t,s))_{ij} \right| ds$ we obtain

$$R_1 = \triangle b^* \left[\left| \frac{\gamma_{24} - \gamma_{25}}{\gamma_{15}\gamma_{24}} \right| \frac{2}{|a_4+k|} + \left| \frac{\gamma_{24}(2\gamma_{35} - a_4+k) + \gamma_{25}(a_4-2a_5+3k)}{\gamma_{32}\gamma_{15}\gamma_{24}} \right| \frac{1}{|a_4+k|} \right.$$
$$\left. + \left| \frac{\gamma_{25}}{\gamma_{15}\gamma_{24}} \right| \frac{1}{|a_5|} + \frac{1}{|\gamma_{15}|} \frac{1}{|a_1-a_2|} \right],$$

$$R_2 = \triangle b^* \left[\frac{1}{|\gamma_{24}|} \frac{2}{|a_4+k|} + \left| \frac{a_4 - 3k + 2a_5}{\gamma_{24}\gamma_{32}} \right| \frac{1}{|a_4+k|} + \frac{1}{|\gamma_{24}|} \frac{1}{|a_5|} \right],$$

$$R_3 = \triangle b^* \frac{1}{|\gamma_{32}|} \frac{2}{|a_4+k|},$$

$$R_4 = 0,$$

$$R_5 = \triangle b^* \left[\frac{2}{|a_4+k|} + \left| \frac{a_4-k}{\gamma_{32}} \right| \frac{1}{|a_4+k|} \right].$$

(24)

Theorem 4. *Let the assumption of Theorem 1 be fulfilled, $(a_4 - k)^2 < 4b$ and the inequality $\max_{1 \leq j \leq 5} \{|R_j|\} < 1$ be true. Then system (16) is exponential stable.*

The proof of Theorem 3 repeats the proof of Theorem 2.

Example 3. *Substituting the values from Remark 1 and setting $k = 1, b = 2$ we obtain*

$$R_1 \leq 133.8894553, \quad R_2 \leq 6.173038374 \cdot 10^{-7}, \quad R_3 \leq 1.511857554, \quad R_4 = 0, \quad R_5 \leq 0.7559286288.$$

The inequality $133.8894553 \cdot \triangle b^* < 1$ implies the inequality $\max_{1 \leq j \leq 5} \{|R_j|\} < 1$.
Thus if $\triangle b^* < 0.7468848071$, then the system (16) is exponentially stable, according to Theorem 4.

5. Influence of Changes in the Right-Hand Side on Behavior of Solutions

Constructing system we neglect the influence of different factors that seem us nonessential. The Cauchy matrix $C(t,s)$ allows us to estimate the influences of all these factors on the solution.

Consider the system
$$Y'(t) - AY(t) = G(t) + \triangle G(t), \qquad (25)$$

where the matrix A is defined by (8) is the matrix of the coefficients of system (7) and $\triangle G(t) \in L_\infty^5$ describes a change of the right-hand side. In the following assertion we estimate the difference between the solution-vector $Y(t) = col\{y_1(t), ..., y_5(t)\}$ of the system (25) and the solution $X(t) = col\{x_1(t), ..., x_5(t)\}$ of the system (7).

Theorem 5. *Under the assumption of Theorem 1 the system (7) is exponentially stable and the following inequality*
$$\|Y(t) - X(t)\| \leq \|C\| \, \|\triangle G(t)\|,$$

is true, where

$$\|C\| = \max_{1 \leq i \leq 5} \left(\sup_{t \geq 0} \int_0^t \sum_{j=0}^5 |c_{ij}(t,s)| \right) ds, \quad \|\triangle G(t)\| = \max_{1 \leq i \leq 5} \operatorname{ess\,sup}_{t \geq 0} |\triangle G_i(t)|,$$

$$\|Y(t) - X(t)\| = \max_{1 \leq i \leq 5} \sup_{t \geq 0} |y_i(t) - x_i(t)|.$$

The proof follows from the representation of solution of system (7).

The estimates of $\|C\|$ can be obtained through the estimates of the elements of the Cauchy matrix obtained in Section 3.

Author Contributions: Conceptualization A.D. and I.V.; methodology, A.D. and I.V.; software, I.V. and M.B.; validation, A.D., I.V. and M.B.; formal analysis, A.D., I.V. and M.B.; investigation, A.D., I.V. and M.B.; resources, I.V. and M.B.; data curating, I.V. and M.B.; writing–original draft preparation, A.D., I.V. and M.B.; writing–review and editing, A.D. and I.V.

Funding: This research received no external funding.

Conflicts of Interest: The authors declare no conflict of interest.

References

1. Marchuk, G.I. Mathematical Modelling of Immune Response in Infection Diseases. In *Mathematics and Its Applications*; Springer: Dordrecht, The Netherlands, 1997.
2. Martsenyuk, V.P. On stability of immune protection model with regard for damage of target organ: The degenerate Liapunov functionals method. *Cybern. Syst. Anal.* **2004**, *40*, 126–136. [CrossRef]
3. Martsenyuk, V.P.; Andrushchak, I.Y.; Gvozdetska, I.S. Qualitative analysis of the antineoplastic immunitty system on the basis of a decision tree. *Cybern. Syst. Anal.* **2015**, *51*, 461–470. [CrossRef]
4. Rusakov, S.V.; Chirkov, M.V. Mathematical model of influence of immuno-therapy on dynamics of immune response. *Probl. Control* **2012**, *6*, 45–50.
5. Rusakov, S.V.; Chirkov, M.V. Identification of parameters and control in mathematical models of immune response. *Russian J. Biomech.* **2014**, *18*, 259–269.
6. Skvortsova, M. Asymptotic Properties of Solutions in Marchuk's Basic Model of Disease. *Funct. Differ. Equ.* **2017**, *24*, 127–135.

7. Belih, L.N. On the Numerical Solution of Models of diseases. In *Mathematical Models in Immunology and Medicine*; Marchuk, G.I., Belih, L.N., Eds.; Elsevier: New York, NY, USA, 1986; pp. 291–297. (In Russian)
8. Bolodurina, I.P.; Lugovskova, Y.P. Mathematical model of management immune system. *Rev. Appl. Ind. Math.* **2008**, *15*, 1043–1044. (In Russian)
9. Chirkov, M.V. Parameter Identification and Control in Mathematical Models of the Immune Response. Ph.D. Thesis, Perm State University, Perm, Russia, 2014.
10. Belih, L.N.; Marchuk, G.I. The qualitative analysis of the simplest mathematical model of an infectious disease. In *Mathematical Models in Immunology and Medicine*; Marchuk, G.I., Ed.; Science: Novosibirsk, Russia, 1982; pp. 5–27. (In Russian)
11. Agarwal, R.P.; Berezansky, L.; Braverman, E.; Domoshnitsky, A. *Nonoscillation Theory of Functional Differential Equations with Applications*; Springer: New York, NY, USA, 2012.
12. Azbelev, N.V.; Maksimov, V.P.; Rahmatullina, L.F. *Introduction to Theory of Linear Functional-Differential Equations*; Advances Series in Mathematical Science and Engineering; World Federation Publishers Company: Atlanta, GA, USA, 1995; Volume 3.
13. Goebel, G.; Munz, U.; Allgower, F. Stabilization of linear systems with distributed input delay. In Proceedings of the 2010 American Control Conference, Baltimore, MD, USA, 30 June–2 July 2010; pp. 5800–5805.
14. Mazenc, F.; Niculescu, S.I.; Bekaik, M. Stabilization of time-varying nonlinear systems with distributed input delay by feedback of plant's state. *IEEE Trans. Autom. Control* **2013**, *58*, 264–269. [CrossRef]
15. Domoshnitsky, A.; Goltser, Y. Approach to study of stability and bifurcation of integro-differential equations. *Math. Comput. Model.* **2002**, *36*, 663–678. [CrossRef]
16. Domoshnitsky, A.; Volinsky, I.; Polonsky, A.; Sitkin, A. Stabilization by delay distributed feedback control. *Math. Model. Nat. Phenom.* **2017**, *7*, 32–45. [CrossRef]

© 2019 by the authors. Licensee MDPI, Basel, Switzerland. This article is an open access article distributed under the terms and conditions of the Creative Commons Attribution (CC BY) license (http://creativecommons.org/licenses/by/4.0/).

MDPI
St. Alban-Anlage 66
4052 Basel
Switzerland
Tel. +41 61 683 77 34
Fax +41 61 302 89 18
www.mdpi.com

Symmetry Editorial Office
E-mail: symmetry@mdpi.com
www.mdpi.com/journal/symmetry

www.ingramcontent.com/pod-product-compliance
Lightning Source LLC
LaVergne TN
LVHW070617100526
838202LV00012B/670